PROCEEDINGS OF SPIE

 SPIE—The International Society for Optical Engineering

Sensor Technology for Soldier Systems

Patrick R. Snow, Jr.
David A. Randall
Chairs/Editors

15 April 1998
Orlando, Florida

Sponsored and Published by
SPIE—The International Society for Optical Engineering

Volume 3394

SPIE is an international technical society dedicated to advancing engineering and scientific
applications of optical, photonic, imaging, electronic, and optoelectronic technologies.

The papers appearing in this book comprise the proceedings of the meeting mentioned on the cover and title page. They reflect the authors' opinions and are published as presented and without change, in the interests of timely dissemination. Their inclusion in this publication does not necessarily constitute endorsement by the editors or by SPIE.

Please use the following format to cite material from this book:
   Author(s), "Title of paper," in *Sensor Technology for Soldier Systems*, Patrick R. Snow, Jr., David A. Randall, Editors, Proceedings of SPIE Vol. 3394, page numbers (1998).

ISSN 0277-786X
ISBN 0-8194-2843-4

Published by
SPIE—The International Society for Optical Engineering
P.O. Box 10, Bellingham, Washington 98227-0010 USA
Telephone 360/676-3290 (Pacific Time) • Fax 360/647-1445

Printed in the United States of America.

Contents

SESSION 1

Sensor Technology for Soldier Systems I

Audio sensors and low-complexity recognition technologies for soldier systems

W. M. Campbell[a], K. T. Assaleh[b], and C. C. Broun[a]

[a] Motorola SSTG, Scottsdale, AZ

[b] Rockwell Semiconductor Systems, Inc., Newport Beach, CA

ABSTRACT

The use of audio sensors for soldier systems is examined in detail. We propose two applications, speech recognition and speaker verification. These technologies provide two functions for a soldier system. First, the soldier has hands-free operation of his system via voice command and control. Second, identity can be verified for the soldier to maintain the security integrity of the soldier system. A low complexity, high accuracy technology based upon averaging a discriminative classifier over time is presented and applied to these processes. For speaker verification of the soldier, the classifier is used in a text-prompted mode to authenticate the identity of the soldier. Once the soldier has been authenticated, the interface can then be navigated via voice commands in a speaker independent manner. By using an artificial neural network structure, a high degree of accuracy can be obtained with low complexity. We show the resulting accuracy of the speaker verification technology. We also detail the simulation of the speech recognition under various noise conditions.

Keywords: Soldier systems, portable computers, speech recognition, speaker verification

1. INTRODUCTION

The recent introduction of portable computers into soldier systems has made possible many new interface options. Through current implementations, a soldier has access to a complete graphical system through a head-mounted display. With such a system, the soldier has detailed control of sensors, radios, and other equipment. Many current implementations use a pointing-device to provide a "point-and-click" approach to navigating the interface. This paper discusses the alternate use of speech recognition and speaker verification technologies and the associated benefits.

Speech recognition has the potential to provide a system which has a more natural interface to the soldier; the soldier navigates and controls system functions using voice commands. Speech recognition has the added bonus that it provides hands-free (and potentially eyes-free) operation of the interface. This ability is critical in many situations. Simple command-and-control speech recognition can be added with extremely low complexity, less than 10 DSP MIPS (million instructions per second). Thus, speech recognition will not overwhelm the soldier system.

Speaker verification technology provides a simplified security method for a soldier system. Speaker verification is the process of authenticating the identity of an individual by his voice.[1] The advantage of using a biometric such as voice over a password-based method is several fold. First, the soldier system can be personalized so that the system can only be used by the intended individual(s). Second, passwords are subject to many problems. If not periodically changed, they are subject to "cracking." Passwords also are difficult to remember in adverse situations. Third, speaker verification is low cost to implement. No additional keypad is required to implement a password; only the microphone used for speech recognition is needed. Verification is implemented using a similar technology to speech recognition.

In this paper, we present a low-complexity method for implementing speech and speaker recognition technologies. Section 2 discusses the basic scenario for speech recognition and the discriminative approach to training and testing. Results and implementation considerations are discussed. Section 3 details methods for applying similar technology to the speaker verification process. We discuss an efficient enrollment and verification scenario. Results are given for the YOHO database.

Other author information: (Send correspondence to W.M.C.)
W.M.C.: E-mail: p27439@email.mot.com
C.C.B.: E-mail: p27611@email.mot.com
K.T.A.: E-mail: khaled.assaleh@rss.rockwell.com

2. SPEECH RECOGNITION

2.1. Basic Scenario

The basic scenario we propose is to use speech recognition for command and control of the soldier system, see Figure 1. The soldier speaks a command in the current context and the system responds. For instance, if the heads-up display is in the radio menu card, then the soldier could speak "power on" to turn the radio on, or "mute" to mute the radio audio. This scenario has been successfully implemented through a proof-of-concept demonstration for the Force XXI Land Warrior project.

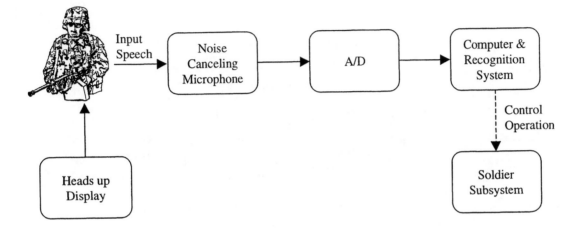

Figure 1. Basic speech recognition scenario.

Recognition is initiated in Figure 1 through a "push-to-talk" interface; the soldier presses a button and then speaks the command. The audio input is through a noise-cancelling microphone. Recording is performed for an amount of time equal to the length of the longest command. This process simplifies the endpointing required and also eliminates the need for keyword spotting. The latter operation would increase power drain since recognition would need to be performed continuously.

The final output of the recognition unit is a control operation to an appropriate subsystem of the soldier unit or a change in the default operation of the computer control.

There are several reasons for implementing only command and control and not more complex continuous speech recognition functions. First, a low computational complexity implementation is desired. Many current soldier systems perform numerous functions using limited power resources (batteries), e.g., communications, image processing and display, video processing, etc. If speech recognition consumes too much CPU time and power, then it is likely to be abandoned. Second, discrete utterance speech recognition is adequate for providing control of the system. Most control operations of the system are via a set of menu cards which can be navigated by speech. Third, command and control provides the critical hands-free operation of the system.

2.2. Classifier structure

The structure of the classification process for speech recognition is shown in Figure 2. First, feature extraction is performed on the discrete-time input speech, $x[n]$. The result of the feature analysis is a sequence of feature vectors, $\mathbf{x}_1, ..., \mathbf{x}_M$. The feature vectors and models of the speech commands for the current menu are then introduced into a pattern classifier. The output of the classifier is the command best matching the input feature vectors.

Feature extraction for the system is performed as shown in Figure 3. Speech is first processed through an IFIR filter structure[2] (for low complexity). The output is then processed using 12th order LP (linear prediction) analysis every 10ms using a frame size of 30ms and a Hamming window. The speech is preemphasized using the filter

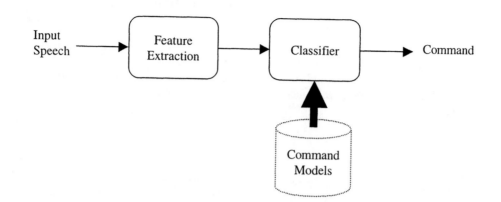

Figure 2. Automatic speech recognition system.

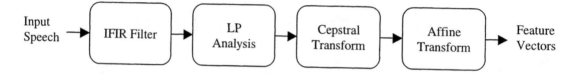

Figure 3. Feature extraction for automatic speech recognition.

$H(z) = 1 - 0.97z^{-1}$ before finding the LP coefficients. The output is passed through the standard LP coefficient to cepstral coefficient transform

$$c_0 = 0, c_n = a_n + \sum_{k=0}^{n-1} \frac{k}{n} c_k a_{n-k} \tag{1}$$

where a_n is the sequence of LP analysis coefficients and c_n is the corresponding set of cepstral coefficients.[3] Note that $a_n = 0$ for n greater than the LP analysis order. Fourteen of the cepstral coefficients, c_1, ..., c_{14}, are formed into a feature vector. Delta-cepstral analysis[4] is also performed if necessary. Delta-cepstral analysis provides critical time information which is necessary for distinguishing confusable words.

Two additional basic processes are performed by the system in feature extraction, channel/noise compensation and endpointing.

Noise and channel compensation are accomplished through the use of an affine transform. Each feature vector, $\mathbf{x}_i$, is passed through an affine transform

$$\mathbf{y}_i = \mathbf{A}\mathbf{x}_i + \mathbf{b}. \tag{2}$$

This approach has proved successful in several situations, see Refs. 5,6. The effect of (2) can be thought of as a two part process. First, the input feature vector is normalized to the channel via $\mathbf{x}$ being mapped to $\mathbf{x} + \mathbf{c}$. This process is performed using cepstral mean subtraction (CMS).[7] Cepstral mean subtraction removes the effects of linear time-invariant processing. In particular, it compensates for different audio paths (particularly the spectral tilt of the microphone input). Second, the translated input is normalized to the training data variance with a diagonal transform, $\mathbf{D}(\mathbf{x} + \mathbf{c})$. This normalization process provides noise compensation. It is well known that the cepstral coefficients shrink towards zero in noise. The normalization process mitigates this effect. The overall compensation, $\mathbf{D}\mathbf{x} + \mathbf{D}\mathbf{c}$ is an affine transform.

Endpointing is the process of detecting which frames of the input correspond to speech input. A simple short-time energy scheme is used. This endpointing method has proved effective for the vocabularies used and has low computational complexity. For each input frame, the short-time energy

$$E_i = \sum_{i=0}^{N_s-1} |x[n]|^2 \tag{3}$$

is calculated, where N_s is the number of samples in the frame. The energy is then broken up into two distributions, one representing the signal and one representing the noise. The average energy level (in dB) is calculated for each of these distributions resulting in $\bar{E}_{\text{noise}}$ and $\bar{E}_{\text{speech}}$. The noise floor is then set at

$$\alpha \bar{E}_{\text{noise}} + (1 - \alpha)\bar{E}_{\text{speech}} \tag{4}$$

where α is typically 0.9. Every frame above the noise floor is kept as speech.

Figure 4 shows the process of scoring for a particular model in the classifier. In general, a model represents the phonetic content of a particular command for the system; e.g., model 1 may represent the word "one" and model 2 represents the word "two." In Figure 4, the feature vectors produced from Figure 3, $\mathbf{y}_1,...,\mathbf{y}_M$, are introduced into the system. A discriminant function[8] is then applied to each feature vector, $\mathbf{y}_i$, using the jth model, $\mathbf{w}_j$, producing a scalar output, $f(\mathbf{y}_i, \mathbf{w}_j)$. The final score for the jth model is then computed

$$s_j = \frac{1}{M} \sum_{i=1}^{M} f(\mathbf{y}_i, \mathbf{w}_j). \tag{5}$$

The selection of the best model j^* matching the input is then found

$$j^* = \text{argmax}_j s_j. \tag{6}$$

We have found that accuracy can be increased by dividing the scores by prior probabilities, see Ref. 9. Typically, in Figure 4, the discriminant function is an artificial neural network or related structure, see Ref. 10. We use a unique artificial neural network structure with low computational complexity features.

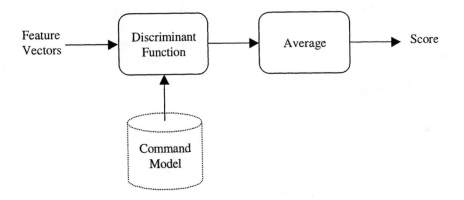

Figure 4. Processing for one model of the classifier.

This classification structure has several advantages. First, the classifier has low computational complexity. For small vocabularies, computing the discriminant function is low cost, and the averaging adds little overhead. In fact, in our implementation, the feature extraction process requires the most computation. Second, the classifier has good discrimination properties. Since the operation of the system is command and control, the designer can influence the selection of the vocabulary. Choosing a non-confusable vocabulary results in good discrimination between the command utterances.

2.3. Training

The system shown in Figure 4 requires discriminative training in order to maximize the recognition performance. There are several considerations for training. First, our proposed system implements speaker independent speech recognition. Speaker independent speech recognition requires that the system have no prior knowledge of the speaker before he uses the system. Therefore, a training database of a "large" sample of the speaker population speaking the commands for the system is needed. Second, the discriminative classifier must be trained using a criterion which maximizes the difference between in-class and out-of-class scores.

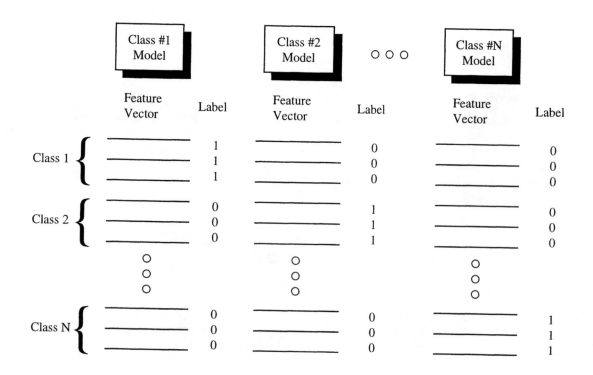

Figure 5. Labeling of feature vectors for training.

The criterion for training that we use is mean-squared error. For each command, utterance, and speaker in the database, the features are extracted. Then, for each command, features are partitioned into in-class and out-of-class sets, see Figure 5. The in-class feature vectors are labeled with a 1; the out-of-class vectors are labeled with a 0. Training is accomplished by finding the model, $\mathbf{w}_i$, that best approximates the labels; i.e., the model, $\mathbf{w}_i$, satisfies

$$\mathbf{w}_i = \mathrm{argmin}_{\mathbf{w}} \sum_{i=1}^{N_t} |f(\mathbf{y}_i, \mathbf{w}) - l_i|^2 \qquad (7)$$

where N_t is the number of training vectors and l_i is the label (either 0 or 1). This criterion is commonly used in the literature, see Ref. 11.

2.4. Results

For an initial implementation, two vocabulary lists were selected, see Table 1. These lists were picked as representative of commands for the soldier system.

A training database of 34 male soldiers was collected. Each soldier repeated each command 10 times. The speech is sampled at 11025 Hz and is 16-bit linear PCM. This database was used for training the system as indicated in Section 2.3. A separate set of 30 male soldiers was collected for testing the system. As in the training database, each command was repeated 10 times. Note that the training condition is quiet; the challenge is for the recognizer to perform well in additive noise with no prior information.

To simulate battlefield conditions, noise was introduced into the testing database to estimate accuracy in several realistic situations. The noise files were obtained from the NOISEX database. Three sources, "leopard", "m109", and "machine gun" were used. To determine the appropriate SNR, these three sources were played in an audio chamber at dBA SPL levels of 114, 100, and 100 respectively, and then recorded with a noise cancelling Gentex microphone while a soldier was commanding the system. Noise was then added to the testing database at the appropriate levels.

Another artificial noise, "colored", was also added to the testing data at 12 dB SNR. This noise was white noise highpass filtered at 200 Hz. The purpose of this noise mix was to test the effect of noise compensation and the robustness of the system at unnaturally high noise levels.

Table 1. Sample vocabulary lists.

List 1	List 2
alert	zero
channel	one
date	two
degrees	tree
display	four
edit	fife
hour	six
location	seven
meters	eight
mile	niner
open	hundred
status	
view	

Table 2 shows the results for the system shown in Figure 2. The number of cepstral coefficients used is 12. In addition, 12 delta-cepstral coefficients are used for the implemented system. Note that the system achieves high accuracy across all realistic situations including impulsive noise such as the "machine gun" and stationary noise such as "leopard" and "m109".

Table 2. Accuracy in percent under different operating conditions using delta-cepstral coefficient features.

Condition	Accuracy List 1	Accuracy List 2
Quiet	95.1	98.0
Leopard Noise	93.6	94.4
M109 Noise	95.4	96.6
Machine Gun	95.1	97.5
Colored	70.3	65.3

For the next set of experiments, all 14 cepstral coefficients are used. In addition, a time feature is added. No delta-cepstral coefficients are used. The time feature is a linear warp of the time axis for the feature vectors. Each speech feature vector is labeled i/M, where $i = 0, ..., M - 1$, and M is the number of speech frames.

Since the classification process averages the score over time, the system lacks information about the order of the feature vectors. Using a time feature provides sequencing information. In addition, since the time feature is only calculated on the speech feature vectors, it is not sensitive to silence in the input. Note, though, that the system is sensitive to the duration of phonetic sounds (but not to the overall duration of the command). This problem is mitigated to some extent by the inherent variability in training data. The results are shown in Table 3. Note that accuracy is improved in all cases even though the number of features is decreased (from 24 down to 15). In particular, the performance in noise is improved.

Further work has been performed with a 100 speaker training database. A complete navigation vocabulary has been designed and implemented.

3. SPEAKER VERIFICATION

3.1. Basic Scenario

The basic scenario for speaker verification is shown in Figure 6. The soldier requests access to the system. The system responds by prompting the soldier to read a particular text. After the soldier has spoken the text, the system either grants or denies access. This scenario encompasses two processes. First, "liveness" testing is performed. By

Table 3. Accuracy under different operating conditions using a time feature.

Condition	Accuracy List 1	Accuracy List 2
Quiet	97.0	98.7
Leopard Noise	96.3	96.4
M109 Noise	96.8	98.2
Machine Gun	97.0	98.8
Colored	78.4	73.5

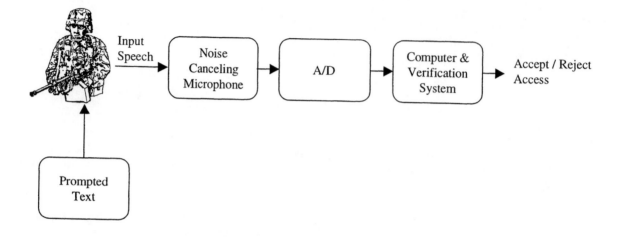

Figure 6. Scenario for speaker verification.

selecting a random set of doublets with which to prompt the user (e.g., "23-45-56"), the system counteracts spoofing (using a tape recorder or other method). Second, after liveness testing has been performed, the system verifies the identity of the individual.

This scenario used for speaker verification is referred to as text-prompted.[1] Other methods of verification include text dependent (e.g., using a fixed "voice password") or text independent (the soldier can utter an arbitrary phrase). There are several advantages of our implementation of text-prompted verification. First, the text corresponding to a "voice-password" does not have to be stored. This feature decreases model size. Second, "liveness" testing is not compromised since text-prompting randomly changes the input. Third, the soldier does not have to remember a "voice password."

3.2. Enrollment

Before verification can be performed, the soldier must enroll in the system. The basic process of enrollment is as follows. First, the soldier repeats a set of phrases as prompted by the verification module. This process can vary from several seconds to several minutes depending upon the security level required by the system. Second, the classifier, shown in Figure 4, must be trained to distinguish between the soldier and other potential impostors.

The process of training the verification system is straightforward. First, since the classifier in Figure 4 is discriminative, we must train the enrollee against a simulated *impostor background*. An imposter background is a fixed set of speakers (typically, for access control, 100 speakers works well) which are used as the out-of-class examples for training. That is, when a soldier enrolls, the model is trained so that the soldier's features are labeled with a 1 and the features from the imposter background set are labeled with a 0. Training is accomplished as shown in Figure 5 and Equation (7).

3.3. Verification

The verification process is shown in Figure 7. As in speech recognition, feature extraction is performed and the result is passed through a classifier. Note that the structure of the classifier in Figure 7 is the same as that of Figure 4. The main difference is that the resulting score is compared to a threshold. If the score is below a threshold T, the claimant is rejected; otherwise, he is accepted. The performance of the verification system is measured in terms of

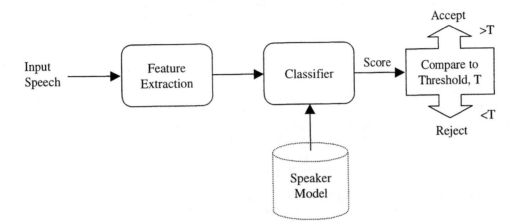

Figure 7. Verification process.

the average EER (equal error rate) across the speaker population. The equal error is determined using the FAR (false acceptance rate) and FRR (false rejection rate). The FAR for a speaker is the percentage of trials that the system falsely accepts an impostor:

$$\text{FAR} = \frac{\text{Number of imposters utterances accepted}}{\text{Number of imposter utterances}} \times 100\%. \tag{8}$$

The FRR for a speaker is the percentage of times access is denied to the correct claimant:

$$\text{FRR} = \frac{\text{Number of speaker utterances rejected}}{\text{Number of speaker utterances}} \times 100\%. \tag{9}$$

The FAR and FRR are dependent upon the threshold T set in Figure 7. The error rate where FAR=FRR is termed the EER for the speaker. The average EER is the average over all speakers in the population

$$\text{Average EER} = \frac{1}{N_{\text{spk}}} \sum_{i=1}^{N_{\text{spk}}} \text{EER}_i \tag{10}$$

where EER_i is the EER for the ith speaker and N_{spk} is the number of speakers tested.

3.4. Results

The verification system was tested using the YOHO database.[12] The YOHO database is a publicly available speaker verification database. The database consists of 138 speakers enrolled in 4 separate sessions. Each session has 24 enrollment combination lock phrases of the form "23-45-56" (3 doublets). For verification, there are 10 sessions consisting of 4 combination lock phrases.

Results are shown in Table 4. The two-phrase test is constructed by concatenating two phrases together in a session. This results in a total of 20 test phrases per speaker. The four-phrase test is constructed by concatenating all four phrases together in a session resulting in a total of 10 tests per speaker.

Several items should be observed about the results in Table 4. First, the average EER appears to approximately halve as the length of the verification phrase doubles. Second, as the amount of enrollment data is doubled, the average EER decreases by 30 to 50 percent. Overall, the table shows that good EER can be achieved with modest complexity.

Table 4. Classifier average EER in percent under different enrollment/verification situations using the YOHO database.

Enrollment Sessions	Two-phrase test	Four-phrase test
1	1.42	0.70
1,2	0.87	0.48
1-4	0.56	0.25

4. CONCLUSIONS

The use of low-complexity recognition technologies for soldier systems has been illustrated for both speaker verification and speech recognition. This technology results in high accuracy (above 96 percent in realistic situations) speech recognition. The recognizer technology was also applied to speaker verification. The resulting average EER's of less than 2 percent showed that the system could provide authentication with a usable level of security. Overall, these technologies provide soldier systems with new advanced functions which dramatically enhance the soldier's capability.

ACKNOWLEDGMENTS

The authors would like to thank Motorola SSTG and the Speech and Signal Processing Lab for support of this effort through Force XXI Land Warrior and IRAD (internal research and development) funding. The work by Khaled Assaleh was performed while he was at Motorola SSTG. The views and conclusions in this document are those of the authors and should not be interpreted as representing the official policies, either expressed or implied, of the U.S. government.

REFERENCES

1. J. P. Campbell, Jr., "Speaker recognition: A tutorial," *Proceedings of the IEEE* **85**, pp. 1437–1462, Sept. 1997.
2. T. Saramäki, Y. Neuvo, and S. K. Mitra, "Design of computationally efficient interpolated FIR filters," *IEEE Trans. Circuits and Systems* **35**, pp. 70–88, Jan. 1988.
3. A. V. Oppenheim and R. W. Schafer, *Discrete-Time Signal Processing*, Prentice-Hall, 1989.
4. F. K. Soong and A. E. Rosenberg, "On the use of instantaneous and transitional spectral information in speaker recognition," in *IEEE Conference on Acoustics, Speech, and Signal Processing*, pp. 877–880, 1986.
5. R. J. Mammone, X. Zhang, and R. Ramachandran, "Robust speaker recognition," *IEEE Signal Processing Magazine* **13**, pp. 58–71, Sept. 1996.
6. Q. Li, S. Parthasarathy, and A. E. Rosenberg, "A fast algorithm for stochastic matching with applications to robust speaker verification," in *International Conference on Acoustics Speech and Signal Processing*, pp. 1543–1546, 1997.
7. B. S. Atal, "Automatic recognition of speakers from their voices," *Proceedings of the IEEE* **64**, pp. 460–475, Apr. 1996.
8. K. Fukunaga, *Introduction to Statistical Pattern Recognition*, Academic Press, 1990.
9. N. Morgan and H. A. Bourlard, *Connectionist Speech Recognition: A Hybrid Approach*, Kluwer Academic Publishers, 1994.
10. S. V. Kosonocky, "Speech recognition using neural networks," in *Modern Methods of Speech Processing*, R. P. Ramachandran and R. J. Mammone, eds., ch. 7, pp. 159–184, Kluwer Academic Publishers, Norwell, Massachusetts, 1995.
11. S. Haykin, *Neural Networks: A Comprehensive Foundation*, IEEE Press, 1994.
12. J. P. Campbell, Jr., "Testing with the YOHO CD-ROM voice verification corpus," in *Proceedings of the International Conference on Acoustics, Speech, and Signal Processing*, pp. 341–344, 1995.

An electronic compass and vertical angle measurement sensor - applications and benefits to the soldier system

Barry Roberts, Angela Johnson, Ron Belt, and Bill Platt

Honeywell Sensor and Guidance Products
2600 Ridgway Pkwy.
Minneapolis, MN 55413

ABSTRACT

The purpose of an electronic *Compass and Vertical Angle Measurement* (CVAM) sensor is to measure the attitude / orientation (i.e., azimuth, elevation, and cant; in terms of true North and local vertical) of the object to which it is attached. This measurement of orientation enables the addition of valuable functionality to the soldier system.

Currently the Force XXI Land Warrior (FXXI LW) program is funding the development of a CVAM and its integration onto the Integrated Helmet Assembly Subassembly (IHAS) of the Land Warrior system. The primary intent of the FXXI LW effort is to produce a *Head Orientation Sensor* (HOS) that, when combined with data from a corresponding *Weapon Orientation Sensor*, enables a new function for the Land Warrior system known as *Rapid Target Acquisition* (RTA).

The RTA functionality is only one of a set of soldier functions enabled by the CVAM. It is the objective of this paper to describe the CVAM/HOS and its various applications within the soldier systems.

The introduction of the paper provides the history and background of the CVAM and its operation. The body of the paper is first focused on an overview of CVAM operation and the means by which it is to be integrated into various systems (to include helmets, advanced weapon sights, chemical sensing platforms, etc.). This will include the unique needs of the sensor such as calibration and parameter adjustment, and its key features that make it excel in its operational environment.

The remainder of the paper focuses on the application of the CVAM to soldier systems; the benefits that CVAM offers to existing functions and the new functionality it enables. Various soldier functions and capabilities (new and pre-existing) are described.

The paper concludes with a description of the status of the effort. This includes the developmental status of the CVAM/HOS, the validation of its design, and its integration into the IHAS and other weapon and sensing platforms.

Keywords: orientation sensing, digital compass, land warrior, integrated helmet system.

INTRODUCTION

Many military and industrial applications require an inexpensive means of sensing the orientation of various objects and platforms. In short, some means of sensing the yaw (azimuth), pitch (elevation), and roll (cant) of the platform is required. A variety of means are available to sense pitch and roll. These range from liquid filled inclinometers to solid-state accelerometers. The performance (e.g., in terms of range of measurement, settling time, temperature dependence) obtainable with the sensors varies greatly. The most demanding applications require $\pm 80°$ in pitch and $\pm 45°$ in roll (and of course, the full 360° in azimuth) accurate to 1 degree within the orientation range with measurements made at a 60 Hz rate.

To sense azimuth involves the use of magnetometer circuitry to measure magnetic North or the use of navigation quality inertial measurement units to perform *gyro-compassing* (which involves the measurement of Earth's rotation). A whole taxonomy of such sensing devices exist, ranging from less than $100 to 10's of thousands of dollars with accuracy ranging from 20 degrees to 0.001 degree. Applications exist for each device within this range of azimuth accuracy.

As will be explored within this paper, the applications range from weapon sights, vehicle compassing, chemical agent detection systems, etc..

Honeywell has developed a *Compass and Vertical Angle Measurement* (CVAM) device under U.S. Army funding as part of the Force XXI Land Warrior (FXXI LW) program[†]. Early work toward today's version of the CVAM was performed under the Integrated Sight Modules program from U.S. Army Night Vision Directorate.

[†] Work was performed as part of the prime contract, number DAK60-94-C-1065, from U.S. Army, SSCOM Natick MA.
(email address for correspondence: broberts@sgp.honeywell.com)

Part of the SPIE Conference on Sensor Technology for Soldier Systems ● Orlando, Florida ● April 1998
SPIE Vol. 3394 ● 0277-786X/98/$10.00

11

The design and development of the CVAM is now complete. It consisted of focused attention to detail that has resulted in a robust sensor that is easily integrated into many different applications. A key element of the successful CVAM integration is the Integrated Process and Product Development (IPPD) process through which the development activity was performed. Members of each sub-team or IPT include representatives from all critically related functions (i.e., Management, Users, Designers, Testing, Specialty Engineering, MANPRINT, Manufacturing). This team is equipped and empowered to make decisions that result in a final integrate-able subsystem or product. The strength of the IPT approach comes from blending, at appropriate times, the various disciplines having a role in the successful development, production, fielding and maintaining of a modular subsystem. The IPTs focus and foremost objective is to provide an integrated product that is affordable and fully utilized by the end user.

Currently the FXXI LW program is funding the development of the CVAM for purposes of *Head (and Weapon) Orientation Sensing* (HOS). Also funded is the integration and test planning required to integrate it into the Integrated Helmet Assembly Subsystem (IHAS) of the Land Warrior system (see [1] for additional information). The purpose of the Honeywell effort is to demonstrate and test potential functional additions to the LW system that are enabled by a HOS and WOS. In particular, the FXXI LW program is funding the development of the software to process the HOS (and WOS) data to provide additional functionality to the warfighter (described later in this paper). The HOS is not currently a part of the LW EMD production system.

It is the CVAM and its soldier system-relevant applications that are the subject of this paper.

1. CVAM DESCRIPTION AND THEORY OF OPERATION

The CVAM is a solid-state, strap-down sensor that measures (and provides in digital form) three angles: magnetic Azimuth (compass heading), pitch (elevation), and roll (cant) angles. The three angles describe the orientation of the sensor (and the platform on which it is attached) relative to Earth magnetic North and local vertical.

The CVAM design uses Honeywell's thin-film, silicon, magneto-resistive (MR) permalloy technology to build the three axis magnetic field sensing elements, and incorporates a tilt sensor, constructed out of a three axis, solid-state, accelerometer package, to eliminate the need for a gimbal. The tilt sensors included in the module provide the orientation information needed for the calibration and computation routines which enable accurate compass heading information at orientations outside of the local horizontal plane. Hence, in addition to providing tilt compensation for the compass, the tilt sensor module also serves as the required means of measuring *vertical* (i.e., pitch) angle and roll angle. The CVAM hardware in its current form, is a small (1.8" x 2.28"), two circuit card assembly with height of less than 0.6". Figure 1 graphically illustrates the hardware and highlights the locations of sensors and connectors.

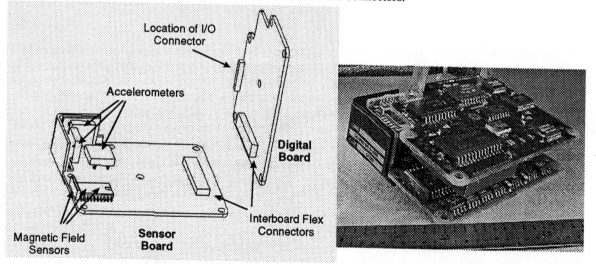

Figure 1: Three-dimensional view of the CVAM (showing sensors and connectors only).

The magnetic field measurements are taken from three separate sensing axes (under the known tilt orientations) and then used to accurately determine the magnetic azimuth / heading of the compass module. Thereafter the magnetic azimuth angle is adjusted for declination to produce true North azimuth. Figure 2 illustrates the coordinate axes of the CVAM and the vectors representing the measured Earth's magnetic field ($\mathbf{H_E}$) and gravity ($\mathbf{G}$) vectors. The figure also contains the geometry behind the computation of the azimuth (ψ), pitch (θ), and roll (ϕ) angles.

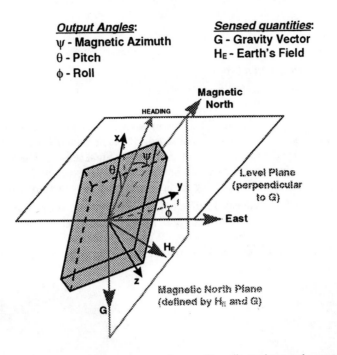

Output Angles:
ψ - Magnetic Azimuth
θ - Pitch
ϕ - Roll

Sensed quantities:
G - Gravity Vector
H_E - Earth's Field

[x,y,z - Coordinate frame of sensor]

Figure 2: The geometry of the CVAM and its measurements.

Operational commands and output data packets of azimuth, vertical, and cant angles are exchanged with the host computer through a serial interface (either RS-422 or RS-232). The CVAM provides azimuth readings to the LW Computer/Radio Subsystem over the range of 0 to 6399 mils (or 0 to 359.9 degrees) and vertical and cant angle measurement readings in the range of ± 1422 mils (or ± 80 degrees). The CVAM module outputs the azimuth, pitch, and roll angle measurements on the digital serial interface at a user selectable data rate of 60 Hz or less. All functions of the CVAM are accessible via its serial interface. For example, the CVAM operating mode (three modes: Run, Idle, Calibrate), the data format, the type of data, and data rate can all be changed by the host computer.

Note that the CVAM must be calibrated for the magnetic environment in which it is placed. For man-portable applications, a closed-form calibration procedure is performed within the CVAM that requires 16 data points to be acquired at various orientations of the CVAM. The 30 second computation time following the data acquisition enables the CVAM to produce accurate azimuth readings over the full range. This calibration is performed only once per configuration and/or mission.

For vehicle-mounted applications, a procedure of external reference alignment of the vehicle and compass is required to account for the "index" / offset error between vehicle and compass. The index error is hence corrected via the software download of the adjustment constants into the nonvolatile memory of the compass. If a "navigation" grade inertial navigation system is available, then the option of using inertial data becomes available, in which the compass measurements can be aligned to the reference of the nav system (which is assumed to be otherwise aligned to the vehicle).

To account for the first and second order disturbances caused by the vehicle's ferrous metals, a Kalman filter (KF) is employed to blend compass data with heading data provided by a inertial navigation system. The KF produces an estimate of the coefficients used in the error model thereby completing the error model for use in correcting CVAM measurements.

Note that all calibration coefficients are stored in non-volatile memory internal to the CVAM. Different sets of calibration coefficients (for when the CVAM is in various environments) can be stored and recalled when the compass is in the corresponding environment.

2. CVAM APPLICATIONS WITHIN THE SOLDIER SYSTEM

There are multiple uses of CVAM data that enable heretofore unavailable LW system functions/capabilities. Such functions include

Rapid Target Acquisition (RTA)

Heading Indicator (HI)

Automatic, Direction-of-View, Map Orientation (MO)

Augmented Reality (AR)

RTA is computer software that operates on the *HOS* and *WOS* data to produce symbology, known as the *Weapon Sight Field of View Indicator* (WSFOVI), that is superimposed on the night sensor's field of view (FOV). The WSFOVI is a visual indication of where the weapon sight's FOV will fall within the FOV of the night sensor. The WSFOVI gives the soldier the situational awareness necessary to enable a *quick* transition between the night sensor and the weapon sight to improve the acquisition and engagement of targets thereby improving soldier lethality and survivability in many scenarios.

RTA operation is illustrated in Figure 3. RTA reduces the time required to switch *from* the night sensor's view of the environment (FOV as large as 40°); shown as the leftmost image in Figure 3), provided by the LW Night Sensor and Display Component (NSDC) *to* the view provided by the Thermal Weapon Sight (TWS) (FOV as small as 3°); the rightmost image in the figure). The reduction in time is measured relative to the time required to perform the same function using the LW system without RTA / HOS.

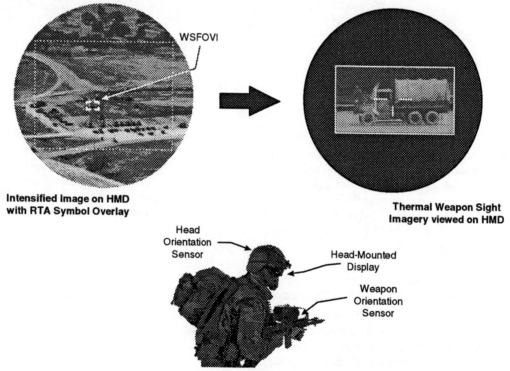

Figure 3: Illustration of RTA usage.

The scenario of usage is as follows. Initially the user will be navigating using the NSDC. A target/area of interest will be identified by the user. The user will slew the weapon such that the WSFOVI will lie on top of the area of interest. With the touch of a weapon-mounted switch the user can then bring the TWS imagery onto the display within the NSDC and have the WSFOVI-designated area contained within the thermal sight's imagery.

Note that the NSDC FOV is 2.6 to 13 times as large as the FOV of the TWS (the TWS has greater magnification). Note also that the NSDC and TWS produce imagery in significantly different wavebands of the EM spectrum. A camouflaged person that is not visible in the NSDC will clearly be visible within the TWS imagery. All of the above, provide the reason for switching between NSDC and TWS and the desire to do so quickly. Without the WSFOVI-aided transition, the basic hand to eye coordination is not sufficient for the user to get the TWS line-of-sight onto the target quickly, hence denying a portion of the benefit of the TWS.

RTA adds no weight to the LW system (but does require the presence of a HOS and WOS in the LW system) and only minimal power (i.e., the computer power required to do the coordinate transforms and computations required to produce the WSFOVI and other supporting symbology).

The *HI* is an indication of the helmet's/head's orientation (relative to magnetic north) that is displayed on the HMD in the form of an annotated *azimuth tape* presented at either the bottom or top of the HMD.

The HI gives the soldier a significant situational awareness advantage. It enables the soldier to take quick compass bearing measurements in the direction of view (i.e., in the viewing direction of head/helmet); providing the means to quickly acquire headings to key landmarks. Also, *indicators* superimposed on the tape indicate the direction to waypoints and landmarks of interest (based on location knowledge obtained via other means, e.g., GPS). What has just been described is graphically presented in Figure 4. An *elevation tape* can also be added along either the right or left sides of the display.

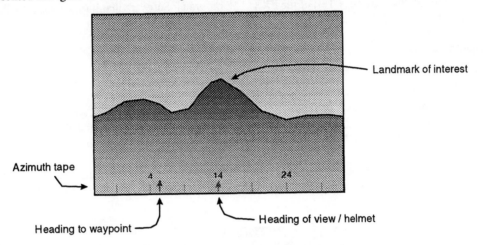

Figure 4: Illustration of the appearance of the *Heading Indicator* function.

Related to the HI application, the *automated rotation of maps* as displayed on the HMD can be implemented once a HOS is present. This would significantly reduce the cognitive workload involved in reading a map and registering map features to world features by rotating the map such that the direction of view is the "up" direction of the map as it is displayed. Without this electronic assist, soldiers are trained to read paper maps by rotating them such that direction of view is "up". Hence, giving the soldier the capability to view digitized maps on the HMD, with the aid of a HOS, is a much more efficient and familiar a way to present map data.

Finally, any means by which the soldier's view of his environment is augmented to improve situational awareness is what is called *Augmented Reality*. The display of symbology that tracks the location of areas of interest (e.g., waypoints, landmarks, etc.) in 3-D relative to soldier location and line-of-sight, falls into this category. For operation in MOUT scenarios in which it is necessary to convey 3-D locations (e.g., bearings to sniper locations on roof tops or windows). A "bulls-eye" of sorts would be presented that would only be centered on the HMD when line-of-sight of the HMD is parallel to the direction of interest. This functionality would be of assistance in quickly coordinating a fire team's attention on a common target in a stealthy manner. Also included in this category are applications involving the tracking of head movement for use in man-machine interface (i.e., controlling and/or interfacing to an electronically controlled system). This would include head trackers (requiring moderate accuracy) used to guide weapon movement. This would also include a means of interfacing to a computer system via head movements as in the HOS being used as the computer's mouse used to control cursor movement for example.

Other applications within the soldier system involves a Weapon Orientation Sensor (WOS). For example, the LW weapon subsystem already incorporates a WOS that is a key part of a target designation functionality. Given that the soldier knows his own location, via GPS for example, the weapon-mounted laser range finder + WOS provide a 3-D vector from the soldier to target. Subsequently, target location can be quickly, automatically, and accurately computed and inserted into a call for fire. Of course, the WOS also supports the RTA functionality (see above). The CVAM is currently part of the Integrated Sight being developed under

Outside of the soldier system the CVAM finds application in sensor platforms and within the navigation systems of vehicles of all kinds. Basically any application that is in need of heading and orientation. Currently, CVAM is being integrated into chemical agent detection sensor and into a military vehicle. Effort is also underway to integrate the CVAM into Unmanned Aerial Vehicles.

3. STATUS AND SUMMARY

The new CVAM design is complete with operational units having been built, tested, and shipped. Integration of a CVAM onto the actual LW helmet and its interface to the Computer/Radio subsystem of the LW system are underway and planned to be completed in time for the Early User Test of the FXXI LW program. Software that implements the RTA and HI functionality have been written, debugged, and tested. Multiple user assessments have been conducted as part of the software development process.

The proof of concept, in-the-field demonstration of the RTA functionality is currently planned by the end of March 1998.

4. ACKNOWLEDGEMENTS

The authors are very thankful for the support, financial and technical, provided by the U.S. Army. In particular the personnel (too numerous to mention) at Natick SSCOM, PM Soldier, PM TSM, the Dismounted Battlespace Battle Lab (Ft. Benning), Night Vision Directorate (Ft. Belvoir), Armament Research Development & Engr. Center (Picatinny Arsenal).

Also the authors are grateful for all of the support and technical interaction with our colleagues at Motorola SSTG (FXXI LW prime contractor), at Raytheon TI Systems (Integrated Sight prime contractor), and at the Honeywell Technology Center.

REFERENCES

[1] B. Roberts, A. Johnson, B. Platt, S. Do, and P. Vetter, "A helmet integration case study: head orientation sensor integration into Land Warrior," *The Design and Integration of Helmet Systems*, December 1997.

Laser Sensor Technology for the 21st Century Soldier System

Jagdish Mathur[a], Wayne Antesberger[b], and Nick Broline[a]

[a]Tracor Aerospace, 6500 Tracor Lane, Austin, TX, 78725
[b]Night Vision & Electronic Sensors Directorate, Fort Belvoir, VA, 22060

ABSTRACT

This paper presents the design and implementation of the laser sensor system for the 21st century soldier under the U.S. Army's Land Warrior program. The system, a low cost, light weight, non-developmental item (NDI), meets the current requirements and provides for future growth to support training and combat identification (ID) functions. This totally integrated laser sensor system has been implemented in an Integrated Product Team (IPT) environment using all commercial off the shelf (COTS) technologies and components. The paper also discuss how the baseline technology for the laser sensor developed under Tracor Aerospace's Independent Research and Development (IRAD) projects was transitioned for the Land Warrior applications.

Keywords: Laser Sensor, Laser Detector, Land Warrior, Soldier System, Digital Battlefield, Combat Identification, MILES Training

1. INTRODUCTION

In the concept of the U.S. Army's Digital Battlefield (DBF), the key to the operation and success is based on sensor systems integrated horizontally and vertically through a digital information network. These sensor systems are the "unblinking" eye of the battlefield, netted together and locked in with a Global Positioning System (GPS) to provide information for situational awareness, Identification of Friend/Foe (IFF), targeting, command and control and all other phases of modern combat. Within this DBF, the 21st century soldier of the U.S. Army will be operating as a self-contained sensor platform[1]. One of the sensor in his suite--- a laser detector--- will perform multiple functions to improve his situational awareness and therefore, increase his survivability to win the battle.

While the laser sensor functions in the battlefield during the 80's were limited to the detection of lethal laser threats, the digital battlefield of the next century will see lasers performing non-lethal functions such as IFF and intelligence gathering. In this new scenario, using the traditional approach of adding sensors in an incremental manner to provide additional capabilities will result in overloading of the soldier. Instead, the sensible and cost -effective approach is to address the requirements in an integrated manner by increasing the functionality of the existing hardware. To accomplish this objective, the U.S. Army's Land Warrior program is taking a revolutionary step by designing a totally integrated laser sensor system for the 21st century digital battlefield.

2. SOLDIER LASER THREATS

The most common threats on the battlefield include lasers operating in the visible to near infrared region of the electromagnetic spectrum. This region, extending from 0.4 micron to about 1.8 micron, has mature laser technology and provides good atmospheric transmission for the propagation of a laser beam. It is most likely that the 21st century Land Warrior will be exposed to a laser threat from this region. As shown in Table 1, a typical threat is a pulsed laser with 1 to 10 MW of peak power, 10 to 50 ns pulse width, and 0.5 to 1.0 mR beam divergence.

Further author information--- J.M (correspondence) e-mail: jmathur@tracor.com; Tel (512) 929-2832
W.A; e-mail: wantesbe@nvl.army.mil; Tel: (703) 704-1728
N.B; e-mail: enb@eng1.tracor.com; Tel: (512) 929- 4314

Part of the SPIE Conference on Sensor Technology for Soldier Systems ● Orlando, Florida ● April 1998
SPIE Vol. 3394 ● 0277-786X/98/$10.00

17

Table 1. Laser Threats and Their Characteristics

Laser Type	Wavelength (μm)	PRF (Hz)	Pulse Energy (mJ)	Pulse Width (ns)
Freq. Double, Nd: YAG	0.53	10-20	100-500	10-20
Ruby	0.694	0.2-2, Single Shot	40-100	30-40
Gallium Arsenide	0.8-0.9	1k-20k	$1-4 \times 10^{-3}$	60-180
Nd:YAG	1.064	1-30, Single Shot	50-200	10-30
Nd:Glass	1.06	0.1-2	50-300	10-50
Erbium	1.54	Single Shot, 0.2-2	50-200	10-50

A soldier standing about a kilometer away from such a laser is exposed to a very wide range of laser power depending upon the wavelength of the laser, atmospheric conditions, and distance from the main beam. For example, Figure 1. shows that a soldier standing 1 km from a Nd:YAG laser of 1MW peak power can be exposed anywhere from 200 W/cm^2 peak power density within 0.5 meter from the main beam on a clear day, to 1 mW/cm^2 within 120 meters from the main beam on a very foggy day. Such variation in power levels demands that the laser sensor has a wide dynamic range and allows a soldier to change its minimum detection threshold to support various Land Warrior missions under different weather conditions. The sensor's ability to adjust its threshold is critical to avoiding unnecessary alerts when no mission-related activity is taking place at the scene. Such alerts could be perceived as false alarms and would degrade the overall utility of the laser sensor.

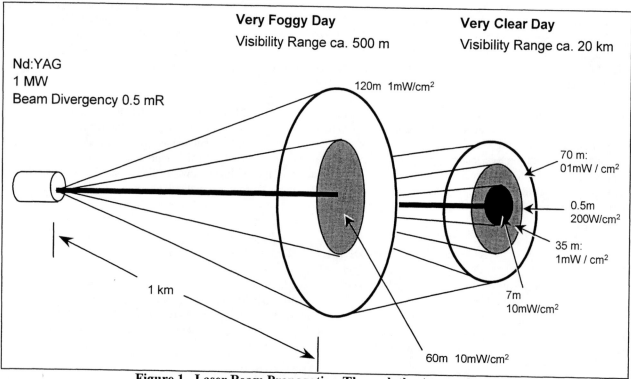

Figure 1. Laser Beam Propagation Through the Atmosphere

3. LAND WARRIOR LASER SENSOR REQUIREMENTS

The Land Warrior program required a very light weight, low risk NDI laser detector to ensure that the soldier receives timely alert of a laser threat. It should be attached to or built into the soldier's clothing and/or equipment without interfering with any combat equipment. The output of the laser sensor should be processed by the Computer and Radio Subsystem (CRS) and the Land Warrior software, and powered by the Land Warrior main battery. An aural notification should be provided through the headset (speaker/microphone) and a visual warning should be provided on the helmet mounted display (HMD). The sensor should provide an alternative method of notifying the soldier of laser detection in the event of computer failure. The other performance specifications are[2]:

- **Spectral Response:** 0.4 to 1.8 micron spectral region to cover the threats listed in Table 1.

- **Detection Sensitivity and Dynamic Range:** Ultimate detection sensitivity of 10^{-5} Watts/cm^2 and not being damaged when illuminated by peak pulsed irradiance levels produced by current battlefield laser sources.

- **Coverage:** Provide uniform coverage to include 360 degrees in azimuth and +/- 45 degrees in elevation relative to horizontal. This coverage be not adversely effected by individual tactical movements.

- **Directional Accuracy:** Indicate quadrant of arrival of laser threats in the required spectral region.

- **Probability of Detection:** Single pulse probability of detection of greater than 90% in all Land Warrior missions.

- **False Alarm Rate:** No more than 0.05/ hour (one false alarm in 20 hours) due to non- optical sources.

- **Power Consumption:** Dissipate less than one (1) watt of power when in operation.

- **Response Time:** Provide alert to soldier within 100 milliseconds

In addition to the above, the requirements ask for complete operational compatibility with the MILES and MILES 2000 system transmitters. An electronically variable detection threshold capability is also required to simultaneously respond to different irradiance levels necessary for various Land Warrior missions. The rationale for this requirement has been presented in the previous section. As discussed further in Section 5., the variable detection threshold feature allows the LDS to accommodate the MILES training and combat ID functions as additional capabilities through a pre-planned product improvement (P3I) process.

4. LAND WARRIOR LASER SENSOR IMPLEMENTATION

The implementation of the Land Warrior Laser Detection Subsystem (LDS) includes all critical issues associated with the development of a laser warning/alert system. These issues, as shown in Figure 2. have been taken into consideration in selecting the design which meets all current requirements and has growth potential to support training and combat ID functions.

Since the Land Warrior is a technology insertion program of the U.S. Army for the individual soldier, the baseline technology for its laser detector subsystem was developed under Tracor's Independent Research and Development (IRAD) program. Through the IRAD program, the company developed two compact laser sensors: Small Light Pulse Alert Receiver (SLIPAR), and Laser Alert Device for Individual Soldier (LADIS). These products validated the basic design and packaging concepts through an extensive testing program and battlefield performance during the Gulf War. Figure 3. shows the Land Warrior laser sensor technology heritage.

The patented technology developed for military application was subsequently licensed to Bel-Tronics-- a high volume commercial electronics company-- with limited rights to manufacture the device called BEL LaserAlert™. This product is for automobile applications and its purpose is to countermeasure speed measuring equipment using lasers.

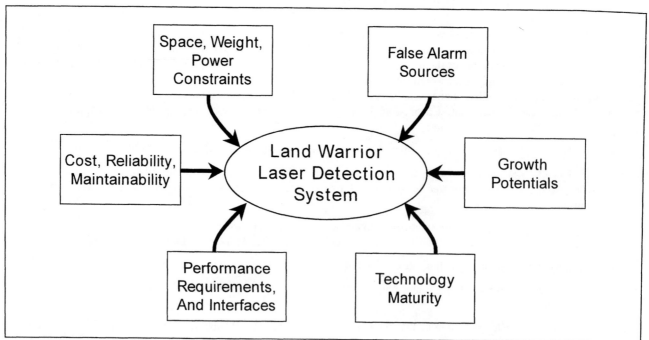

Figure 2. Critical Issues in Land Warrior Laser Detection System

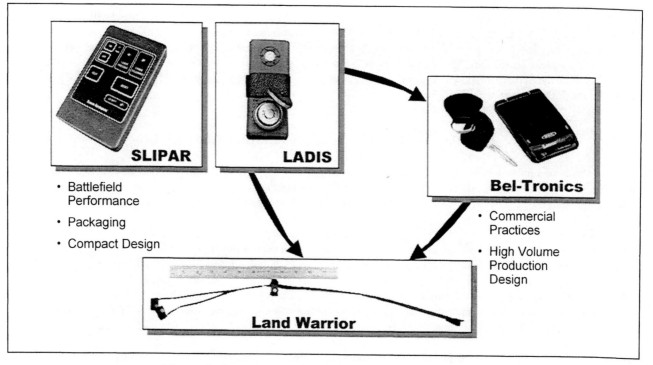

Figure 3. Laser Detection System Technology Heritage

This partnership with Bel-Tronics required Tracor to modify its basic laser sensor design for low cost, high volume production using commercial practices. The experience and the lessons learned through the performance of SLIPAR and LADIS in combat and working with Bel-Tronics on low-cost, high volume commercial manufacturing have been carried forward into the Land Warrior program

The Land Warrior's LDS is a part of the Integrated Helmet Assembly Subsystem (IHAS). It has two subassemblies. The first, as shown in Figure 4., is an integrated array consisting of four detector assembles, each includes a photodiode package with a silicon and an InGaAs detector with a preamplifier. The second, an analog processing module with four channels, each containing Tracor's proprietary laser detection/discrimination circuits, a laser event marker, and a circuit to mark the power received by each detector.

The analog processing module is an integrated circuit component in the Display Control Module (DCM) of the Land Warrior system. Figure 5. shows the analog processing module with the DCM housing. The DCM module is contained in the Electronics Components Housing (ECH) and is a part of the load carrying equipment of the soldier.

The spectral coverage (0.4 to 1.8 μm) is achieved by mating the performance of a silicon detector die with an InGaAs die in a standard TO-5 package prepared by EG&G Optoelectronics, Canada. Both elements provide an unobstructed conical field of view (FOV) of greater than 140 degrees to meet the azimuth and elevation coverage requirements with sufficient overlap to avoid any "blind spots" while maintaining sufficient uniform sensitivity.

The detector array, containing the four photodiode packages with built-in preamplifiers is mounted into the helmet using a field replaceable approach. The decision to locate the array on the helmet was made as a result of a trade off study to determine the best location.

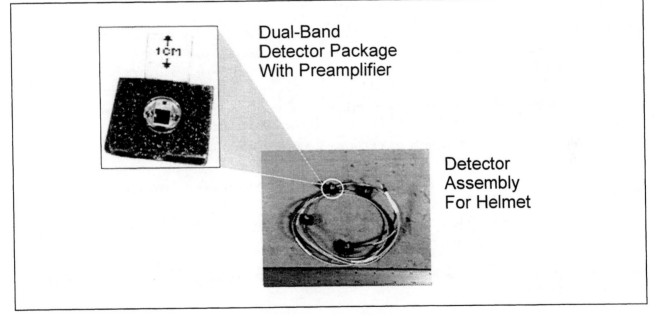

Figure 4. Integrated Laser Detector Array

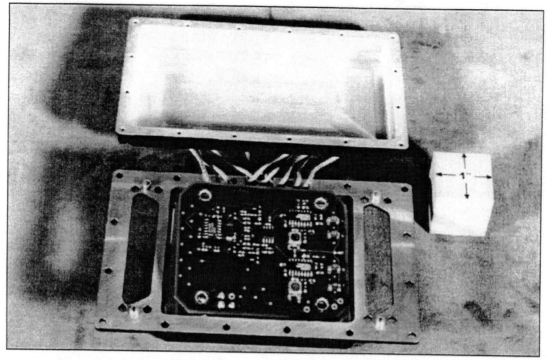

Figure 5. Analog Processing Module with DCM

The helmet was selected because:

♦ Provides greater probability of detection of laser threat than any other location on a soldier's body in an operational environment due to exposure and lesser potential for obscuration from a soldier's clothing or equipment

♦ Easy to attain overlapping field of view to avoid any "blind spots" where a laser event cannot be detected

♦ Allows for a rigid mount that maintains each sensor's orientation relative to the other to provide accurate quadrant of arrival information

♦ Helmet is an essential item in a combat environment, and is always worn by a soldier as compared to other personal clothing items

♦ Ease in integration with the rest of system using the cable path already in place for the HMD

♦ Does not interfere with other combat activities and maneuvers

Figure 6. shows the placement of the laser detectors about the perimeter of the helmet that utilizes the canvas cover as the principle means of integration.

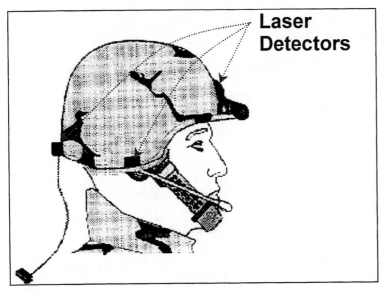

Laser Detectors

The canvas cover, equipped with reinforced "button hole" apertures located at quadrant intervals about the helmet, tightly holds the small casting containing the detector assemblies. A small Velcro patch is used to secure the assembly. The integration approach allows a soldier to replace the detector array in the field following any physical damage. The array is electrically integrated to the inside of the helmet using a single connector/cable carrying power and analog signals. These analog signals have been pre-amplified at the detector module to raise the voltage level and provide first-order reduction in electromagnetic interference susceptibility.

Figure 6. Soldier with Laser Detectors Mounted on Helmet

The pre-amplified analog signal are fed into the analog processing module through a cable where the Tracor's proprietary laser detection technique exploits the exceedingly rapid rise time characteristic of laser pulses to discriminate between lasers and all other optical sources. This technique, tested in several laboratory and field environments, provides discretion against potential false alarm sources found in a military environment and described in Table 2.

The detection is performed by comparing the instantaneous energy levels of each detector assembly with a long-term average background level. An analog signal proportional to the log power on each detector assembly (four in total), and a digital pulse marking a laser event on the occurrence of each laser pulse received above detection threshold, are sent to the CRS for processing. Figure 7 shows the layout of the laser sensor and its interface with the CRS.

Table 2. Battlefield Optical False Alarm Sources

Item	Spectrum	Pulse Nature	Level	Distinguishing Feature
Sun Glint	Broad	Continuous, Related to the Mechanical Motion of Platform	Moderate	Spectrum, Pulse Period
Flares, Fires Rocket Plume, Lamps	Broad	Continuous	Moderate/High	Spectrum, Lack of Period
Gunfire	Broad	Millisecond-Poorly Formed-Aperiodic to 50 Hz	Low	Spectrum, Pulse Shape
Explosive, Sparklers	Broad	Millisecond-Aperiodic or Very Slow	Moderate/High	Spectrum, Lack of Period
Mechanical Chopped Lamps	Broad	10's to 100's of Millisecond, Poorly Formed-10-100 Hz	High	Spectrum, Pulse Shape
E.O. Modulated Lamps	Broad	Microseconds and Up, Very Well Formed- 0.1 to MHz	High	Spectrum
Flash Lamp	Broad	Microseconds-1 Hz to MHz	High	Spectrum, Pulse Length
Sea Glint	Broad	Millisecond and Up	High	Spectrum, Pulse Length

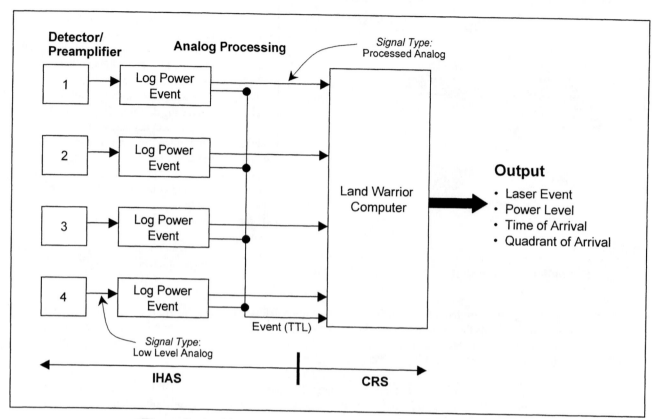

Figure 7. Architecture of Laser Sensor And Interface with CRS

Digital processing performs comparisons of the power from each detector to determine direction of arrival and power level estimates to aid in threat evaluation. The quadrant of arrival is reported for the greatest response in power level. A graphic visual display and an auditory tone provide easy to interpret threat information. This interface and processing approach supports current and future demands on processing loads. The total power consumption for the Land Warrior's LDS is less than 0.5 watts-- one-half of the allocated budget.

In addition to the power consumption, size and weight represent the other two areas of challenge in the Land Warrior implementation. As illustrated in Table3., these challenges have been met by using the highest level of miniaturization possible compatible with cost. Innovative packaging and leveraging on technologies developed and refined in the consumer electronics market made substantial contributions toward meeting these goals.

Table 3. Size and Weight of the Detector Package and Analog Processing Card

Item	Dimensions (cm.)	Weight (grams)
Dual-Band Detector Package with Preamplifier	2.0 x 1.5 x 0.5	3.66
Analog Processing Card	7.3 x 5.8	26.38

5. GROWTH OF LAND WARRIOR LASER SENSOR SYSTEM

The Land Warrior laser detection subsystem is designed to have complete operational compatibility with the MILES and MILES 2000 system transmitters for training purposes. Also, the design has a well defined growth path to perform functions of a laser detector/receiver of the combat ID system. The following paragraphs provide a technical discussion on the growth path.

5.1 MILES Compatibility

Close inspection of the MILES training system discloses that there is an inherent compatibility between the current MILES and MILES 2000 training systems and the Land Warrior's LDS. Table 4. shows the compatibility between the Land Warrior and MILES requirements.   These compatibilities are further discussed in the following sections:

Table 4. Land Warrior and MILES System Parameters

Parameter	Land Warrior	MILES
Spectral Response	0.4 to 1.8 μm	0.88 μm
Sensitivity	10 μ w	10 μ w
Field of View	360^0	4Π sr
Pulse Width	10-500 ns	100 ns
PRF	> 50 KHz	3 KHz (48 KHz timing)
Response Time	100 ms	N/A
Power Level Discrimination	Yes	N/A
Post-Detection Processing	Not Required.	Word Decode

5.1.1   Spectral Coverage and Pulse Processing

The silicon detector provided in the Land Warrior laser detection system to cover the visible and near IR band is capable of providing adequate sensitivity to match the MILES sensor standard at the GaAs ($\lambda = 0.88$ μ m) wavelength. And since the LDS operates on a "pulse by pulse" basis, it is capable of responding to pulse repetition frequency (PRF) exceeding 50 K pulses per second. Therefore, the LDS has the ability to detect and process pulse streams which are encoded in terms of pulse-to-pulse timing or in "pulse present/pulse absent" (binary) format, or in combinations of the two, such as in the MILES system.

The ability to process fast pulses in the "front end" of the Land Warrior's LDS places a potential burden on the follow-on digital processing. Measures can be provided in the digital processing chain to time tag laser events to aid in decoding coded pulse streams with a resolution of 10 to 20 ns. This resolution far exceeds the requirements for MILES and future MILES coding requirements which demand a 48 KHz. (20.8 microsecond period) sample rate to capture all MILES code types[*]. In summary, there is no intrinsic limitation in the laser detection function, including the analog processing (filtering and logarithmic conversion) element, which would preclude the extraction of coded information contained in a laser pulse stream with a high degree of timing accuracy.

---

[1][*]Although the MILES code has a 3 KHz. pulse rate, the basic code is expanded by offsetting of the pulse position in fine increments from those established by the basic 3 KHz. pulse rate.

5.1.2   Detector Placement and Integration Requirements

MILES detection techniques use detectors placed over the soldiers body for spatial resolution. However, the Land Warrior system is helmet mounted in its current configuration,. Therefore, additional detectors must be provided to sense "hits" to other parts of the soldier. These additional detector assemblies, similar to the helmet mounted, can be integrated as an appliqué to the baseline system, and would require only the silicon detector die. In order to provide access to existing processing paths and power without disturbing the baseline design, the integration must occur at one of the connector sites in the laser detection subsystem-to-CRS cable path. Once the electrical interface is established, there are no constraints as to the location and configuration of the add-on detector assemblies. This approach would provide no additional burden to the baseline Land Warrior soldier in non-training applications, and only modest loading during training exercises.

5.1.3 MILES System Solution and Land Warrior

Our analysis indicates that the MILES and Land Warrior laser detection functions would operate compatibly and simultaneously with no conflict. Two discriminants can be used to separate the MILES detection function from the laser threat warning function: expected power level and PRF. The MILES function is characterized by a moderate (3 KHz.) PRF coded pulse train and is low in power. In contract, the PRF of a battlefield laser threat is much lower (nominally 100 Hz. maximum), and the power expected from a direct illumination or a near-main beam miss is much greater than a MILES illumination. A combination of these two discriminants will separate the functions adequately; and therefore, the laser warning functions and the MILES function can be met simultaneously with the same hardware. We believe the addition of MILES functions in the current design of the Land Warrior, as described here, will most likely result in a *decrease* in the soldier's burden compared to a stand alone MILES added on the top of his personal clothing and individual equipment.

5.2 Combat ID Requirements.

A key requirement for a laser interrogator based combat ID Systems for soldier is to use a eyesafe laser (e.g., 1.54 micron region) to avoid an eye injury. This spectral band is covered by the Land Warrior's LDS in its current design. The other key requirement is the ability to interrogate a *specific* soldier, using a coded laser pulse, without unintentionally interrogating other soldiers outside a specific area. This specific area, or the "zone of interrogation", is the function of the combat ID system's maximum interrogation range in meter (m), and its interrogation angular resolution in milliradians (mR). A very sensitive laser sensor system, operating at its full sensitivity, will not be able to limit its response within the zone of interrogation without a built-in power discrimination/reporting capabilities. Since the LDS is capable of reporting power level of a laser pulse over a wide dynamic range, it can support power level discrimination in concert with the digital processing of the interrogating pulse to perform the combat ID functions. The PRF response of the Land Warrior's LDS far exceeds the frequency of coded pulse produced by an interrogator in a combat ID system.

As discussed above, the LDS is capable of providing power discrimination and process coded interrogating pulses. These features allow it to grow into and support MILES and the combat ID functions without significant changes in its baseline design. Like the MILES system, the impact of integrating the combat ID functions in the current design of the Land Warrior will most likely result in a *decrease* in the soldier's burden as compared to a stand alone approach requiring additional hardware borne by the soldier. However, this comparison does not include a laser interrogator carried as a separate item or added to his gun.

6. CONCLUSIONS

In summary, the paper describes the design and implementation of the laser detection subsystem for the Land Warrior program with emphasis on total integration and identification of a road map for its future growth. Since the system is completely integrated, the implementation includes partitioning the design between the IHAS and CRS to avoid duplication and redundancy. The complete integration required the two IPTs to work on a regular basis and make all critical design decisions as one team. The location of the laser detectors on the Land Warrior helmet is based on careful tradeoff studies and analysis to assure complete coverage against present and future battlefield laser threats.

as the use of a MILES applique'. The integrated design approach implemented in the program will *reduce* the current estimate of the soldier's burden when the combination of MILES and combat ID are brought together as one system-- sharing the Land Warrior's laser detector assemblies, its power supply, and its computer resources. Obviously, there will be a corresponding decrease in program and system cost as the development, integration, and production of stand alone detectors and supporting hardware/software items are not duplicated.

ACKNOWLEDGEMENTS

This work was supported by the Land Warrior program under the U.S. Army Contract DAAB07-95-C-H302. The authors wish to thank members of the Integrated Helmet Assembly Subsystem (IHAS) and the Computer Radio Subsystem (CRS) IPT team members for their technical discussion during the design and implementation of the laser detectors. Constant encouragement and leadership support of Bob Whitcraft of Honeywell (IPT Leader, IHAS) contributed significantly towards accomplishing the objectives of our efforts. Finally, a special note of thanks to EG&G Optoelectronics, Canada for their professional commitment and timely response in providing the dual- band photodiode packages.

REFERENCES

1. Marquet, L.C., and J.A. Ratches, " Sensor Systems for the Digital Battlefield", *Digitization of the Battlefield II,* Proceedings of SPIE, Vol 3080, pp 6-15, 1997

2. " Land Warrior System Specifications" - Specification A3246133M- Draft"

Three Combat ID Systems for the Dismounted Soldier

Charles Brenner Ph.D. and Kevin Sherman

Motorola
Space and Systems Technology Group
Communication Systems Division / Warrior Technology Section
Scottsdale, AZ 85252

ABSTRACT

Three different Combat ID systems for the Dismounted Soldier are described and areas of commonality and differences are noted. CIDDS (Combat ID for Dismounted Soldiers) is a complete Stand-Alone system for non-Land Warrior equipped soldier's. To the weapon, it adds a laser interrogator that produces a Multiple Integrated Laser Engagement Systems (MILES) like pulse coded waveform and a RF receiver. To the soldiers helmet, it adds four laser detectors and a RF transmitter that uses frequency diversity and TDMA to respond. Land Warrior Combat ID, being developed on the Force XXI Land Warrior program, adds a similar laser interrogator to the weapon but uses existing laser detectors on the helmet. Detection of a laser interrogation automatically causes a RF reply via the Land Warrior Soldier Radio (LWSR) that operates in the 1.8 GHz frequency range using a TDMA format. Helicopter to Dismounted Soldier ID (HDSID) uses the SIngle Channel Ground Airborne Radio System (SINCGARS) SIP+ radios, operating in the 30 to 88 MHz band, for both interrogation and reply. The SINCGARS SIP+ ground receivers interrupt normal voice or data transmission/reception every second for a very short interval and switch to receive on the "Interrogation" net. Up to three helicopters can make interrogations during any one-second interval. CIDDS and Land Warrior CID are presently in development, while HDSID is a recently completed study.

Keywords: Combat ID, Land Warrior, Laser interrogator

1. INTRODUCTION

Statistics of dismounted friendly fire occurrence in actual combat are not well documented. However, live simulated field engagements at the Army Combat Training Centers reflect an unacceptably high level of fratricide of dismounted soldiers. In particular, exercises conducted at the Joint Readiness Training Center (JRTC), which involve a substantial level of dismounted infantry forces, show that almost half of dismounted friendly fire was from direct fire sources. The large majority of that friendly fire was caused by small arms and machine gun fire. An effective method of soldier-to-soldier combat identification has the potential to significantly reduce the occurrence of friendly fire and to increase combat effectiveness, including improved situational awareness.

Combat ID for Dismounted Soldiers (CIDDS) and Land Warrior-integrated Combat ID (LW-CID) will provide effective soldier-to-soldier CID by providing the following performance: Probability of correct ID of 0.975 for prone-to-prone soldiers at 500 meters, and probability of correct ID of 0.95 for standing-to-standing soldiers at 1100 meters. The probability of false ID shall be no greater than 0.01. CIDDS and LW-CID are expected to be used in a variety of different scenarios such as grassy/wooded terrain and urban areas, but always where a clear line-of-sight is achieved since any opaque obstacle blocks the laser interrogation. The weapon-mounted portion of the system shall operate on M16 and M4 family of weapons, the M240, M-249 and M60 machine guns, and be compatible with Land Warrior weapon systems.

Reductions in dismounted soldier friendly fire from rotary wing platforms can also be achieved using Helicopter to Dismounted Soldier ID (HDSID). Range objectives for HDSID are a probability of correct ID of 0.95 at a range of greater than 7 km. This range represents the maximum effective firing range of many helicopter weapons.

In providing soldier-to-soldier CID, developers must impose several constraints, primarily minimizing added weight to the soldiers helmet and weapon. In addition, it is desired to integrate training functionality with the CID system. Man-portable

28

Part of the SPIE Conference on Sensor Technology for Soldier Systems • Orlando, Florida • April 1998
SPIE Vol. 3394 • 0277-786X/98/$10.00

CID equipment should also have battery life specified in weeks or months as opposed to hours and days. Interchangeable operation should be achieved between primary and rechargeable batteries.

The Stand-Alone, Land Warrior-integrated and helicopter CID systems are described in the next three sections.

2. STAND-ALONE CID CAPABILITY - CIDDS

The Combat ID for Dismounted Soldiers program was awarded to Motorola in July 1997. Lockheed-Martin Information Systems is a subcontractor for the Weapon Unit (laser interrogator) and helmet mounted optical detectors. Raytheon Defense Systems, El Segundo (formerly Hughes) is a subcontractor for the Weapon Interface/Mounting. Initial Army test and evaluation of CIDDS is planned for the summer of 1999.

The system uses a narrow beamwidth laser mounted on the weapon as the interrogation device. Laser operation at 1.54 microns makes it invisible to the naked eye. Four photodetectors and four antennas are contained in a CIDDS helmet cover that is stretched over a standard Personnel Armor System, Ground Troops (PASGT) helmet. When laser interrogation pulses are incident on one or more of the photodetectors, an interrogation is detected. When this occurs, a RF transmitter on the helmet responds with an omni-directional RF reply. When this reply is received by a RF receiver integral to the CIDDS weapon unit, a "friend" indication alerts the interrogating soldier.

Figure 1 shows a pictorial diagram of a CIDDS equipped soldier interrogating a "friendly" CIDDS equipped soldier. The CIDDS weapon unit, containing a laser interrogator and RF receiver, is mounted above the gun barrel and is boresighted to the weapon. The soldier activates the weapon unit through a switch mounted near the non-firing hand. This initiates an encoded laser interrogation. The laser interrogation is received and decoded by the CIDDS equipped target and triggers an encoded RF response on frequencies prescribed by the interrogation message. The interrogated soldier is notified of the interrogation. The interrogator's unit receives and decodes the RF response to the laser interrogation and notifies the soldier of a friendly response. In the event that the target is not CIDDS or LW-CID equipped, the soldier will receive no response to the interrogation.

The CIDDS system provides other functionality and a system architecture that allows for future growth. CIDDS contains an IR aim light comparable to the AN/PAQ-4 IR aim light. CIDDS also provides a Stand-Alone Training Engagement Simulation (TES) capability through the use of the CID detectors as MILES detectors and replicating the MILES Small Arms Transmitter (SAT) functions in the weapon unit. The laser assembly is designed with space for a fourth laser, which could provide day visible aim light as a growth capability. An RS-232 port provides an interface with other equipment, such as the LW system, a CIDDS programming station, or potentially with other weapon-mounted sensors.

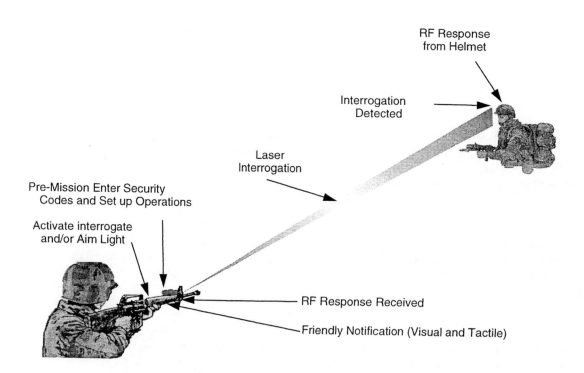

Fig 1 Soldier with Stand-Alone CIDDS Weapon Unit Interrogating a "Friendly" CIDDS Equipped Soldier

Fig 2 Detailed Rear View of Stand-Alone CIDDS Weapon Unit

Figure 2 shows a conceptual rear view of the weapon unit. On the left and top are two flat discs that are used for boresight alignment of the laser assembly. The set of four pushbuttons (O, E, Left arrow and Right arrow) are used for scrolling/entering data on the two-row Liquid Crystal Display (LCD) display. On the right side is a tightening screw used for clamping the CIDDS weapon unit to the weapon rail. The small circles to the left and right of pushbuttons are indicator lights. The two connector jacks below the LCD display are for connection of the interrogation/aimlight switch assembly and a RS-232 cable.

Figure 3 shows a conceptual view of a CIDDS helmet. This view shows two of the four antennas and two of the four small photodetectors mounted just below the antennas. The RF transmitter is mounted on the top of the helmet.

Fig 3 Stand-Alone CIDDS Helmet Unit

2.1 Weapon Interrogation Laser Transmission

The Stand-Alone CIDDS laser uses a beamwidth of approximately 10 mrad. This is wider than a MILES laser which allows an Interrogating soldier to get a positive "friend" respond without requiring "perfect aim" alignment on the target. The optical beam also has sharp cutoff on the edges to minimize the number of "friends" within the interrogated cross-section when a group of soldiers are in close proximity. The Stand-Alone CID system uses a pulse train that is similar to the MILES code. This allows the use of a MILES-like decoder for both the CID and MILES functions. Each frame is transmitted multiple times.

A continuous series of repetitive frames is called an interrogation message. Repetitive frames improve the laser detection probability. Each CID interrogation message contains a CID code, an interrogator ID and an encoded RF channel. Interrogation messages are repeated continuously, with each cycle specifying a different RF channel. The cycle is stopped when either the soldier releases the interrogate switch or a "friend" response is received. Additional interrogations can be performed by releasing and reactivating the interrogate switch. Use of different codes by CIDDS in the interrogation process provides security against an adversary who might copy the CID interrogation and attempt to produce a false CID interrogation. The interrogation ID can be changed periodically with the periodicity dependent on clock time accuracy of the users. For example, by changing the code every x hours, a user time uncertainty of up to +/- x hours is allowed.

A low power wireless link is used from the weapon unit to the helmet unit. This link is used to send system configuration information loaded into the weapon unit by the user. This data transmission is performed whenever new system settings are required. The system design provides for flexibility in updates.

A RS-232 port provides four functions: 1) Rapid loading of configuration information from the CIDDS Programming Station; 2) Periodic software version updates and readout of important stored parameters, 3) Interfacing with the Land Warrior system as described in Section 3, and 4) interface to auxiliary weapon sights.

2.2 Helmet Optical Detection and RF CID Response

A helmet cover that contains laser detectors, a RF transmitter and four antennas is fitted over the helmet. Each antenna provides 90 degrees of azimuth coverage, so that the set of four provides total 360-degree coverage. The elevation pattern of the antenna is also approximately 90 degrees. Four photodetectors are placed in quadrant locations around the circumference of the helmet. These photodetectors are used for detection of either a CID interrogation at 1.54 microns or detection of a MILES weapon fire hit at 0.904 microns. The pulse detection timing clock is asynchronous with the many possible interrogator clocks. Therefore the MILES proven Boolean OR combination of two successive frames is used to reduce the probability of missing a pulse detection due to optical beam scintillation. When a CID interrogation is detected, the Interrogator ID and RF channel is decoded. This initiates a RF response on the specified RF channel.

Once a CID interrogation is detected, the RF transmitter is turned on and tuned to the specified RF channel. Initial activation power up time delay of the RF transmitter causes the RF response to be delayed for a duration of one or two interrogation messages. Then RF response transmissions begin on the last received RF channel number that was decoded from the laser interrogation message. The transmitter first transmits a response message from one pair of diametrically opposed antennas and then repeats the same response message from the other pair of diametrically opposed antennas. Each response message includes the coded versions of the Interrogator ID number and Responder ID number.

Switching of antenna pairs essentially eliminates the interferometric lobing that occurs when two more coherent sources radiate with overlapping beam patterns. The pair of RF response messages are transmitted using a Time Division Multiple Access (TDMA) format which reduces the probability of RF response collisions when two or more "friend" soldiers are illuminated simultaneously. This can occur when two or more soldiers are in close group or when soldiers are collinear with the interrogation beam (with varying ranges).

2.3 Weapon RF Detection of CID Response

Initiation of a CID interrogation turns on the RF receiver. The RF receiver is tuned to the RF channel used in the current laser interrogation message. It is tuned to a new, randomly selected RF channel shortly after the beginning of each new laser interrogation message. This assures that the receiver is always tuned to the RF channel that the "friend" RF response used for transmission. Once a "friend" response has been received, the laser interrogation is automatically terminated. Otherwise it remains on until the interrogation switch is no longer depressed. Automatic shutoff minimizes the amount of time RF is being transmitted by the "friend" who was interrogated. Detection of a "friend" response, causes an LED indicator to blink ON for a short duration on the weapon unit and/or the interrogate switch to vibrate. The range of the system is enhanced by the use of a directional antenna on the weapon receiver. The beamwidth is not important, since whenever a successful laser interrogation occurs, the weapon is always pointed at the target, which corresponds to the peak of the beam.

Use of Interrogator and Responder ID numbers eliminates the possibility of an interrogator receiving a "friend" response from himself due to close-in stray laser reflections from items such as close-in tree branches or specular type reflectors.

2.4 Stand-Alone CIDDS Summary

CIDDS will provide the dismounted soldier with the potential reduction of the incidence of friendly fire and increased situational awareness. It also combines functionality that has been identified as being required by nearly all future dismounted soldiers: combat identification, IR aiming and TES training. .

3. LAND WARRIOR-INTEGRATED COMBAT ID – LW-CID

Force XXI Land Warrior is a Science & Technology program, prime contracted by Motorola SSTG, developing new technologies for insertion into the Land Warrior Engineering Manufacturing Development (EMD) system (Raytheon is the prime contractor Land Warrior EMD). One of these enhancements is an integrated combat ID system, or Land Warrior Combat ID (LW-CID).

The LW-CID system uses the same laser interrogator unit (CIDDS Weapon Unit) that mounts on the weapon as the Stand-Alone CIDDS system (with some software modifications to allow communication to the Land Warrior computer via the RS-232 interface). Thus the laser message is identical between the two systems. The Land Warrior being interrogated uses the existing four photodetectors that are mounted in the Land Warrior helmet cover to sense the laser interrogation. The helmet cover has a wiring harness that goes between the helmet and the Computer Radio Subsystem (CRS) via a Display Control Module (DCM), which converts the analog pulse signals to binary pulses of approximately the same width. Functionally, detection of the laser burst occurs exactly as in the Stand-Alone CIDDS system, providing: 1) CID code; 2) RF channel; and 3) Interrogator ID number. The difference is the RF response is now transmitted using the existing LW Soldier Radio. At the interrogating Soldier, reception is accomplished using the Soldier Radio as the receiver. When a CID-equipped Land Warrior initiates a CID interrogation, the Soldier Radio is automatically interrupted and forced to a CID receive mode. The RF channel is sent from the weapon unit to the CRS by a hard wire connection. Upon reception of the RF response, the LW CRS processes the response data and alerts the interrogator of the "friend" response. The alert can be provided at the weapon in the same manner as the CIDDS system, with additional methods also available, such as a message presented in the helmet-mounted display or with an audible tone or synthesized voice message in the helmet speakers. Each of these alert forms is under the soldier's control. The LW-CID system is completely interoperable with Stand-Alone CIDDS. Figure 5 shows a block diagram of the complete system. This block diagram shows the simplicity of the interfacing connections between the weapon unit and helmet to the soldier's backpack that contains the Computer Radio Subsystem.

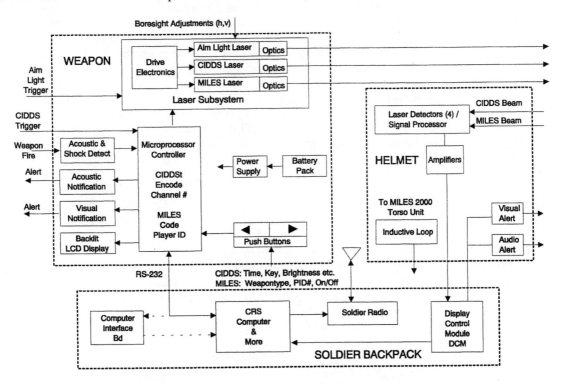

Fig 5 Block Diagram of the Integrated LW/FXXI CID System

3.1 Weapon Interrogation Laser Transmission

The Integrated FXXI LW-CID system uses the exact same laser pulse train as Stand-Alone CIDDS. Minor differences are the receiver signals cannot be directly fed to the receiver but now are sent via the RS-232 link from the weapon to the CRS. These signals are: 1) Interrogation switch activation/deactivation; and 2) RF channel number. Related enhancements planned include replacing the wired connection between the weapon and CRS with a wireless interface, (Enhanced Weapon Interface – under development by Battelle Memorial Institute) so that the weapon is not tethered to the soldier.

3.2 Helmet Optical Detection and Soldier Radio RF CID Response

The LW helmet contains four laser detectors that are electrically connected to the CRS. This allows pulse detections to be fed to the CRS where the pulse decoding is performed on a small new card, which is added to perform CID and some other enhancement functions. Initial interrogation detection either interrupts ongoing RF transmissions or turns on the RF transmitter. The RF channel received is sent to the Land Warrior Soldier Radio. The Soldier Radio operates in the frequency band from 1755 to 1850 MHz and transmits a peak power of 1 watt. During normal Land Warrior radio communications, the Soldier radio uses a TDMA time slot format. When a CID interrogation is detected, it generates a priority interrupt, which causes the radio to begin operation in the CID mode. This mode transmits the same modulation format as the Stand-Alone CIDDS helmet transmitter. This allows interoperability between the Stand-Alone CIDDS and the LW-integrated CID systems. While the frequency channels and the modulation are identical, there are differences in the antennas used. The Soldier Radio uses an omni-directional, vertically polarized monopole antenna rather than the four patch antennas. Thus there is no need for switching between orthogonal pairs of patch antennas that occurs with the CIDDS system. Therefore, the same signal is repeated twice to produce a similar modulation envelope. With the same interrogation laser data, the transmitted RF response message is identical in frequency, bit rate and message structure in both systems. The RF message repeats the interrogator ID number and inserts the responder ID number.

3.3 Soldier Radio RF Detection of CID Response

Initiation of a CID interrogation forces the interrogating soldier's radio to RF receive ON mode. The receiver is tuned to the RF channel used in the current laser interrogation message. It is tuned to a new channel shortly after the next laser interrogation message begins. This assures that the receiver is always tuned to the RF channel the "friend" RF response is transmitted on. The RF channel is sent from the weapon unit to the CRS using an existing RS-232 port on both the weapon unit and the CRS. Once a "friend" response has been received, the laser interrogation can be automatically terminated. Otherwise it remains on until the interrogation switch is no longer depressed. Automatic shutoff is desirable since it minimizes the amount of time RF is being transmitted by the "friend" who was interrogated. Detection of a "friend" response causes an LED indicator to blink ON for a short duration on the weapon unit and/or the interrogate switch to vibrate. The CRS can also present a "friend" message in the helmet-mounted display or with an audible tone or synthesized voice message in the helmet speakers. Each of these alert forms is under the soldier's control

3.4 LW-CID Summary

The approach taken to integrate combat ID into Land Warrior utilizes several functions of the existing LW EMD system, including:
   - Soldier Radio transmitter to respond to laser interrogations
   - Soldier Radio receiver at the Interrogator receives the RF response
   - Pentium computer subsystem performs CID processing
   - Helmet-mounted photodetectors detect CID and MILES laser transmissions
   - Helmet mounted display and speakers provide "friend" indications

The LW-CID approach also uses the CIDDS weapon interrogation unit. This facilitates interoperability with the CIDDS system. As in the Stand-Alone system it also allows elimination of a separate AN/PAQ-4 Aim Light and the need for an entire MILES system (a torso component is still required). This results in fewer components and reduced cost. The most significant factor is no additional weight on the helmet.

4. HELICOPTER TO DISMOUNTED SOLDIER IDENTIFICATION CAPABILITY – HDSID

ITT Aerospace, a long-time prime contractor for SINCGARS radios, developed the SIP+ system for CID between helicopters and SINCGARS-equipped ground vehicles. The SIP+ CID system uses existing SINCGARS radios and accomplishes the CID function by only adding software modifications. This system, which operates in the frequency band from 30 to 88 MHz, was successfully tested at Yuma Proving Grounds in 1997. RACAL is developing a SINCGARS-compatible radio for use in the Land Warrior EMD system, which is called the Squad Radio. Motorola was funded by the government to study CID operation between helicopters and dismounted soldiers using the SIP+ system. The objective was to determine the best approach to integrate the SIP+ combat identification function onto the Land Warrior system.

The SIP+ approach is different from the two previous CID systems in that: 1) RF transmission is used for both interrogation and response; 2) GPS position and time is required at both the helicopter and the dismounted soldiers; and 3) all the SIP+ equipped soldiers within a user defined area/radius of the target aimpoint respond when interrogated. Figure 6 shows a typical scenario with two helicopters interrogating and two groups of soldiers on the ground. The number of soldiers responding is reduced by transmitting an aimpoint and a radius centered on the aimpoint. Only those soldiers who are located within the circle respond to the interrogation.

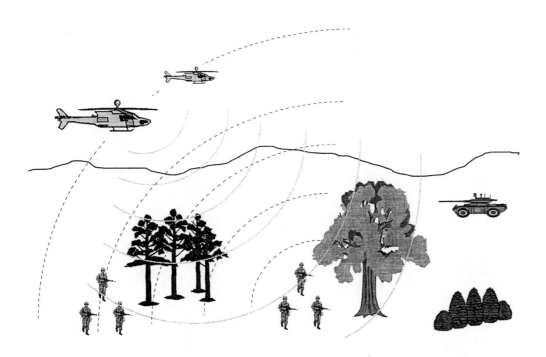

Fig 6   Scenario of Helicopters Interrogating Dismounted soldiers

4.2 HDSID System Operation Using Three Independent Nets: Reservation, Interrogation and Response

As mentioned before the SIP+ CID system uses existing SINCGARS radios and accomplishes the CID function by only adding software modifications. These additions add three independent networks: Reservation, Interrogation and Response. Some minor hardware additions are required in the Squad radio and hardware interfacing to display and controls to the helicopter are required.

The first network used in an interrogation is the helicopter radio Reservation Network. This network is used to prioritize and organize interrogations when multiple helicopters transmit a reservation request during the same one-second time interval. Each helicopter desiring to interrogate transmits a reservation request and then switches its receiver to the reservation net and listens for other reservation requests during the remainder of the listen interval. Near the end of the one-second interval, if a helicopter received three other reservation requests, it would transmit an order message in order slot number -3. Other helicopters receiving this message would then recognize this helicopter as the "lead" helicopter. Figure 7 shows a typical reservation scenario in which Interrogator #1 hears two other reservation requests and since it heard nothing in order slot –3, then transmits its order message in slot –2 and becomes the lead interrogator. Interrogator #2, which received one other reservation request, plans to transmit on order slot –1, but cancels it since it receives a message in order slot –2. From this procedure, a lead helicopter is automatically determined. If a helicopter making a reservation hears no other requests, then it knows that it is the only helicopter that desires to interrogate. In this case, there is no need to transmit an order.

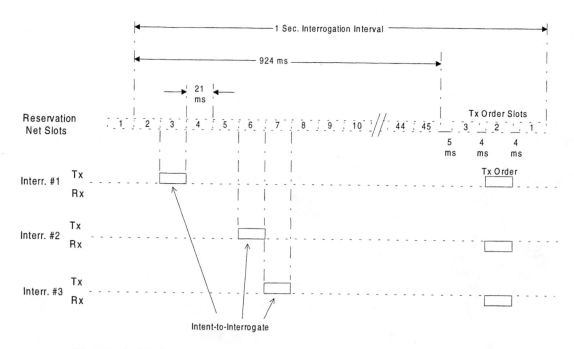

Fig 7 Typical Helicopter Reservation Net Messaging Activity with 3 Interrogation Requests in 1 Second

During the next one-second interval, when there is more than one reservation request, the lead helicopter transmits an "interrogation order" to the other helicopters that designates the interrogation order during this interval. Each interrogating helicopter then make two interrogations, one at the beginning of the interval and a second interrogation, approximately 1/3 of a second later. The second interrogation is made because in some cases the ground receivers could be performing a non-interruptible synchronizing operation. The second interrogation allows reception of interrogations since no synchronization interrupts occur at this time. Reception of the interrogation message occurs since each ground radio interrupts its normal voice or data once every second for a very short interval and switches to the "Interrogation" net, which has its own unique frequency hopping code. This results in a small increase in bit error rate due to the error correcting codes used for digital voice and data. Figure 8 shows a typical interrogation time sequence for the multiple interrogations.

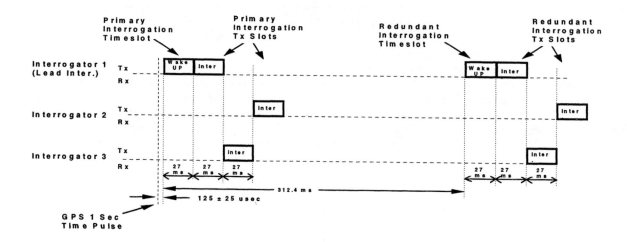

Fig 8 Typical Helicopter Interrogation Net Messaging with 3 Interrogations

After the second interrogation sequence, the Dismounted Soldier/ground radio switches to a separate "Response" net used to minimize collisions when several ground soldiers respond. The Response Net uses a TDMA time slot system with 32 slots. Each responder responds in four randomly selected slots from the 32 possible. For a successful "friend" response to be received, only one response with no collision overlap is required. This TDMA structure effectively reduces probability of no response due to collisions to very near zero. A typical random response TDMA sequence in response to three interrogations is shown in Fig 9.

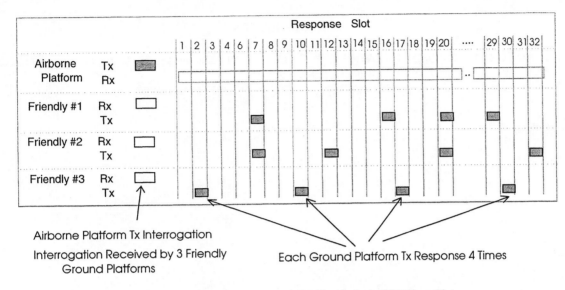

Fig 9   Responders Transmit Randomly in 4 of 32 Time Slots

4.3 Helicopter to Dismounted Soldier CID Range Performance

Range predictions are based on a helicopter at an altitude of 60 m and soldier antenna at a height of 1.5 m.
      Uplink interrogation response range: 3 - 4 km
      Downlink interrogation response range: 8 - 11 km
The uplink response range is reduced compared to the downlink because the helicopter has a transmitter power of 50 watts while the LW Squad Radio has a transmitter of 2 watts.

4.4 HDSID Summary

One advantage of this HDSID approach is that existing hardware can be used to implement SIP+ for dismounted soldiers. In addition, SIP+ can potentially reduce the incidence of friendly fire from another source besides other dismounts, that being rotary wing platforms. However, the difference in response range vs. interrogation range makes the operational effectiveness indeterminate. Theoretical range calculations show that additional uplink power is needed.

5. CONCLUSIONS

Combat Identification for the dismounted soldier, one of the U.S. Army's greatest assets, is becoming a reality. It is being provided to the present force structure in the form of CIDDS and to the dismounted soldiers of the 21st century with Land Warrior Combat ID. The systems will be interoperable, will increase situational awareness and increase survivability for the dismounted soldier. In addition, the Army has begun investigation into providing combat ID of dismounted soldiers for air platforms.

6. ACKNOWLEDGEMENTS

This work described was supported by the following U.S. Army organizations: PM-CID and Natick RDEC under contracts for CIDDS, FXXI LW and HDSID. The authors wish to thank the professionals within those organizations. Additional information for this paper was also provided by Mark Doerning of Lockheed-Martin Information Systems, Phil King of ITT Aerospace/Communications, and Paul Kempin of Motorola SSTG.

7. REFERENCES

1. STANDARD FOR MILES COMMUNICATION CODE STRUCTURE, document PMT 90-002, US Army STRICOM, AMSTI-SI, 12350 Research Parkway, Orlando, FL 32826-3276
2. Combat Identification for Dismounted Soldiers RFP, DAAB07-97-R-S852 (12 May 1997)

SESSION 2

Sensor Technology for Soldier Systems II

Technology Advanced Mini-Eyesafe Rangefinder (TAMER)

Glen P. Abousleman and Bill Smeed

Motorola, Space and Systems Technology Group
8201 E. McDowell Road
Mail Drop H1176N
Scottsdale, AZ 85257
(602) 441-2193
glen1@cwlab.geg.mot.com

Abstract

The Technology Advanced Mini-Eyesafe Rangefinder (TAMER) module is a portable, lightweight (7 lbs), hand-held, target determination system. This rapid prototype program involved the integration of a Motorola 68360 microprocessor, electronic compass, laser range finder, GPS, 4 PCMCIA expansion slots, 0.7-inch micro display, digital camera, floating point unit, and various communications interfaces.

The CPU computes an absolute target position based on laser range to target, C/VAM azimuth and inclination inputs, and absolute GPS position. This target position is automatically formatted into a standard military surveillance report and stored in local non-volatile memory. The operator can attach to a SINCGARS radio or to any RS232 compatible system (e.g., cellular telephone).

To facilitate the above functionality, the TAMER system utilizes various power saving strategies including software-geared power reduction, power supply configuration, external device integration, and incorporation of low-power ICs.

Additionally, TAMER utilizes state-of-the-art digital image compression technology. This custom image coder is based on wavelet decomposition and trellis-coded quantization (TCQ). The algorithm enables TAMER to transmit useful imagery over its severely disadvantaged wireless link.

Keywords: TAMER, laser, rangefinder, target determination, GPS, image compression

1. Introduction

The Technology Advanced Mini-Eyesafe Rangefinder (TAMER) system is a battlefield sensor/communications device developed at Motorola, Space and Systems Technology Group. The TAMER is a lightweight (~7 lbs.), man-portable, integrated unit which allows a soldier to range on a target and determine its absolute position, capture a digital image of the object, and transmit this data via a Single Channel Ground and Airborne Radio System (SINCGARS) radio or cellular phone, to a command post or main base. A number of applications have been identified for this device, including reconnaissance, surveillance, battle damage assessment, and call for fire. The TAMER is an integration of a number of technologies such as a high-performance processor, a laser rangefinder (Mini Eyesafe Laser Infrared Observation Set-MELIOS), a digital compass, a coarse acquisition code Global Positioning System (GPS) multichip module [1] (1.3" x 1.3"- the world's smallest GPS receiver), and a P/Y code external interface. PCMCIA expandability allows further incorporation of a VGA microdisplay in the eyepiece of the system, a cellular modem, a digital camera, and memory expansion. The Active Matrix Electroluminescent (AMEL) [2] microdisplay allows users to view maps, images, and menus laid over the field of view in the eyepiece. The TAMER unit has been fielded at various military exercises for the Army, Marines, National Guard, and Special Forces.

The TAMER system also has an interface with the Distributed Interactive Simulation (DIS) system that is used by the Army to train soldiers and evaluate military equipment. The DIS system consists of a local area network (LAN) which connects various virtual and real equipment to a simulator (SUN workstation), to a three-dimensional visualization tool

40

Part of the SPIE Conference on Sensor Technology for Soldier Systems ● Orlando, Florida ● April 1998
SPIE Vol. 3394 ● 0277-786X/98/$10.00

(Silicon Graphics workstation), and to an interface which communicates with various battlefield devices. The simulator generates battlefield scenarios in detail, and the visualization tool enables the user to look into a particular scenario.

The latest TAMER development activity was incorporation of remote control capability. A soldier can set up a remote-controlled TAMER in the field, and control the unit and perform surveillance in a safer environment. This would entail transmitting image and information data over a wireless link, such as a radio or cellular phone, to a remote operator who would be controlling the TAMER from a graphical user interface (GUI) program based on a laptop computer. Furthermore, Motorola investigated incorporating night vision, field-of-view magnification, and additional low-power enhancements to prolong battery life.

2. The TAMER

The TAMER contract was structured to encompass four phases of development. The following sections describe each of these phases, and the lessons carried over to each succeeding phase.

2.1 Phase I

The goal of the first phase of TAMER was to build a prototype system as rapidly as possible. The objective was to get actual equipment into the user's hands as quickly as possible so that valuable feedback could be obtained. Time spent on peripheral issues would only reduce the effectiveness of building a prototype system. This prototype system had to meet the following specifications:

- Exist in proper form factor
- Integrate with MELIOS, including extraction of laser range and compass
- Integrate GPS
- Compute target position
- Transmit target position in a standard format
- Function in the field

Following this plan, the system was built from concept to proper form factor and fielded within 11 weeks. The functionality included the following:

- 8-bit HC11 running at 3.6 MHz
- 32K ROM
- 8K RAM
- Internal Motorola Oncore GPS receiver
- Three 9-pin serial connectors for external communications
- Conduit connecting MELIOS data lines with TAMER
- PCMCIA slots (nonfunctioning)
- Linear DC-DC power regulator, ~30% efficient
- Approximate operating time: 10 minutes

Within 11 weeks, a unit was available to be fielded at the National Training Center (NTC) in Ft. Irwin, California. The first unit was highly successful in accomplishing its mission of providing the soldiers with a unit capable of being used in the soldier's environment. This was evidenced by the trials at NTC in Ft. Irwin, California. The unit was utilized for nearly a year in various trials and trade shows, and gave the user community a view of what TAMER could provide. Two trials were conducted at NTC with the 2nd armored division stationed at Ft. Hood. Another trial was conducted at Ft. Hood for the Force XXI Operation 256 exercises.

Peripheral options were excluded from the prototype. Design efficiency and quality assurance were also early casualties. The unit suffered from difficult assembly procedures, and power-saving issues were shelved. By maintaining primary focus, however, a system was constructed in a very short time period that satisfied the goals set forth in Phase I [3].

2.2 Phase II

With Phase I successfully completed, a prototype TAMER system now existed that would be used extensively to gather user feedback. Phase II endeavored to expand upon the capabilities of Phase I by including additional peripherals that would enable Phase II to serve as a basis for Phase III and Phase IV. The following features characterized the Phase II TAMER:

- 25 MHz 32-bit CPU
- 2M RAM
- 1M EEPROM
- Integrated menu display
- SINCGARS capability
- 6-button interface

- External Precision Lightweight GPS Receiver (PLGR) (P/Y signal) capabilities
- Power enhancements
- Switching DC-DC power regulator, ~85% efficient
- 4 PCMCIA slots (nonfunctioning)
- Approximate operating time: 1 hour

Phase II involved a complete redesign of the TAMER hardware. The Motorola MC68360 32-bit microcontroller was selected as the main computational engine. Advantages of this processor as compared with the HC11 were numerous, including [4]:

- 32-bit addressing and data
- Integrated low power modes
- Integrated Communication Processor Module (CPM)
    - Provides 6 Universal Asynchronous Receiver Transmitter (UART) capable TTL level communications ports
    - Removes need for external UARTs, reducing valuable real estate consumption
- 25 MHz operation
    - Provides on-the-fly frequency throttling either through internal phase-locked loop or internal clock divider
    - Allows extreme power savings by allowing CPU to run at slow speeds at low load
- Software configurable
    - Allows nearly all options to be fully software configurable, many of them on-the-fly
    - Provides superior software flexibility in configuring the microcontroller
- Large memory map
    - Allows many separate functions to be memory mapped into the CPU memory map
    - Provides eight internal chip selects to remove real estate-consuming "glue" logic

The Electrically Erasable Programmable Read Only Memory (EEPROM), consisting of two 512KB X 16-bit EEPROMs configured to provide 1 MB of 32-bit EEPROM, provided a firm base on which to build. With erasing capabilities, new software could be added to the system without burning additional PROMs. This greatly reduced cycle time. When utilized in conjunction with one of the three serial ports, the ability to download code into the system, using nothing but simple serial communications, provided great flexibility. This flexible platform simplified software upgrades both for the engineering team as well as for end users.

PLGR communications protocols incorporated within the TAMER gave it the ability to acquire GPS location using the superior P/Y GPS signals. The PLGR connects to the TAMER through one of the three external RS-232 ports. PLGR communication protocol software extracted time/position data from the PLGR, as well as Figure-of-Merit (FOM). Upon receiving an acceptable FOM, the TAMER conserved power by terminating communication with the PLGR [5], [6].

The above enhancements not only provided a much better development base, they also provided superior power management. Even with very little effort devoted to this issue, TAMER's operational life increased from ten minutes to one hour.

2.3 Phase III
In Phase III, modularity was incorporated into the TAMER by using PCMCIA slots. In summary, the following features characterized the Phase III TAMER:

- 68360 25 MHz 32-bit CPU
- 2M RAM
- 1M EEPROM
- 8/16 bit PCMCIA capability

- Custom PCMCIA Software Drivers
- PCMCIA Cellular Modem
- PCMCIA AMEL Video Card
- PCMCIA Digital Camera

- PCMCIA SRAM Card
- AMEL Microdisplay (640 X 480 X 64 monochrome)
- Microdisplay field-of-view (FOV) overlay optics
- Baseline TIFF 6.0 Graphic File Reader

- GPS Datum Support
- GPS Lat-Long/MGRS conversion rework
- Software power consumption studies and enhancements

Phase III dealt in part with the addition of PCMCIA expansion slots and associated peripheral work. These slots provided significant flexibility in hardware design. Off-the-shelf PCMCIA cards can be purchased and used with the system. The system needs only a software upgrade to interface with a new PCMCIA card.

Phase III also involved adding several new features to the system through the now-functioning PCMCIA slots. The PCMCIA expansion proved to be one of the more controversial additions to the TAMER system. User feedback ranged from seeing no purpose for the slots, to providing great utility. It should be noted, however, that since PCMCIA expansion capabilities were introduced into TAMER, other dismounted soldier development projects have also incorporated the functionality and expandability that PCMCIA slots provide.

The TAMER is powered by a Motorola MC68360 with a custom operating system. As a result, off-the-shelf (OTS) software is not generally available to run on the TAMER. Accordingly, software would need to be hardware independent, and available in source code to be compiled with the TAMER compiler. Additionally, PCMCIA drivers must be written specifically for the TAMER platform so that the desired PCMCIA card can be used with the system.

The PCMCIA cellular modem provided the first wireless capabilities of the TAMER, apart from SINCGARS radio support [7], [8]. Motivated by discussions with the Marines, the cellular option allowed for a smaller total hardware package because of the smaller size of a cell phone as compared to a SINCGARS. Use of this technology allows position information to be sent remotely, and also allows commands to be transmitted to the TAMER, thus enabling full remote-controlled operation.

The PCMCIA AMEL video card was also developed on the contract. This type II card (length: 85.6mm, width: 54.0mm, thickness: 5.0mm) incorporated the driver circuitry necessary to control the Planar AMEL 640 X 480 X 64 monochrome microdisplay. This microdisplay is 0.7" along the diagonal. A large area of read-write VRAM served as the video buffer. This memory region was memory-mapped into the TAMER addressing region so that the individual pixels could be controlled by direct writes to memory. This simple scheme allowed rapid integration of the AMEL video card with the TAMER.

The PCMCIA digital camera card was an excellent fit into the TAMER system. The digital camera, produced by VVL, had dimensions of 312 X 287 pixels, with 8 bits of radiometric quantization. The PCMCIA digital camera used a 16-bit interface which was not compatible with the TAMER PCMCIA configuration [9], [10]. The TAMER PCMCIA slots were 8/16 bit slots with Intel byte ordering. Unfortunately, the TAMER system used Motorola byte ordering. With an 8-bit card, byte ordering is not an issue because only one byte is utilized. With a 16-bit card, however, the bytes are reversed. In addition, the bytes are aligned differently in the 32-bit data path. Accordingly, special conversion hardware was developed to allow the 16-bit cards to communicate appropriately with the TAMER system. With these issues addressed, the camera provided digital imagery that could be displayed on the TAMER AMEL display. This imagery could also be compressed and transmitted.

The PCMCIA SRAM card support was also added to the system to demonstrate the benefits of mission-specific customization. In one configuration, the SRAM card was used to store a TIFF graphics image and display it on the AMEL microdisplay. A PCMCIA SRAM card can be utilized in a manner that would allow individual TAMER systems to be customized by simply plugging in the appropriate card. By utilizing specific data formats, the TAMER can scan a SRAM card upon its detection. If it detects that the card is programmed with customization data, this data can be downloaded immediately. Customization data can include maps of the region, or intelligence and mission data. Supplemental software required for the mission can also be supplied, and would allow the TAMER to perform additional tasks. This can all be performed without any modification to the base-line hardware or base-line software of the TAMER system. All that is required is that the SRAM card be programmed and distributed to each TAMER before a mission. Removal of the card from the TAMER system would return it to its standard configuration. This type of end-user customization can significantly increase the useful life of the base-line TAMER equipment.

The microdisplay, because of its small size (0.7" diagonal), can be configured as an overlay projection. Optics were designed to perform this task. The microdisplay was mounted above the eye fixture, and the image was projected into the TAMER field-of-view. These optics allowed light transmission of 80%. The overlay application allowed the user to view menus, maps, or other information, while simultaneously observing the current field-of-view of the TAMER.

Power utilization was also investigated during this phase. With the additional peripherals being added to the system, and the overall desire to reduce size and weight, power consumption was an important consideration. Using the software configurable 68360 system clock frequency, an analysis was made to determine power consumption along the spectrum of available clock frequencies. Significant power savings were observed when lower clock frequencies were utilized. The analysis and subsequent actions regarding power utilization will be discussed in following sections.

2.4 Phase IV

Phase IV was an aggressive phase that involved a re-design of the major circuit boards inside the TAMER. The final version of TAMER is shown in Figure 1. In summary, the following features characterized the Phase IV TAMER:

- MC68360 33 MHz 32-bit CPU
- MC68882 33 MHz 32-bit FPU
- Wavelet Image Compression
- Distributed Interactive Simulation (DIS)
- Remote Control of TAMER
- PC GUI Interface

- Pan-Tilt Module
- VersaPoint Input Interface
- Differential GPS
- Integrated Chassis
- Exploration of 3.3V parts
- Night Vision and Optics Magnification Explorations

The major focus of Phase IV was the pursuit of remote control operation of the TAMER. Remote operation can significantly increase the number of scenarios in which a TAMER can contribute to the overall mission success. Remote operation can allow a TAMER to be set up in a remote location and left unattended for long periods of time. Communications can be maintained via wireless link. By dialing into the TAMER through the cellular communications link, a TAMER can be ordered to take pictures, determine target position, transmit data, and change its orientation through the use of a pan-and-tilt mechanism that was also incorporated during this phase.

Other options include mounting a TAMER to a vehicle such as a HMMWV. This option allows users to remain in an enclosed environment while controlling the TAMER. A physical link may be used in this situation. The TAMER would now be able to perform image gathering and target determination. Additionally, while tapped into the HMMWV power system, the TAMER would have a virtually unlimited source of power, and could operate in a fully optimal state.

The desire to use TAMER in remote situations required the addition of three major features to the Phase III TAMER. The first of these was the addition of a communications protocol which would allow the TAMER to communicate. The second was the creation of an external GUI interface that would control the TAMER from a remote site. The third was the pan-and-tilt system which would allow the TAMER to change its orientation after deployment.

A wireless link is always subject to transmission errors, and thus, any communication protocol utilized must be able to detect and possibly correct those transmission errors. The communication protocol developed for TAMER allowed commands and data to be transmitted using packets. Most commands require only a few bytes. Data transfers were handled in blocks using larger packets. A 16-bit cyclic redundancy check (CRC) was made after each packet transmission, and a retransmit was ordered if the CRC failed. The protocol would adjust packet sizes based on the number of transmission errors. The greater the error rate, the smaller the packet.

The PC-based GUI was developed in the Windows 95 environment. The GUI design was implemented in a control-panel-type of implementation. TAMER orientations were shown using compass icons. The latest TAMER image was displayed in a dedicated window. The latest TAMER data, such as TAMER location and target location, were also displayed on-screen. The GUI was designed to run either through a direct physical link, or through a cellular link. GUI development followed a rapid prototype methodology whereby feedback could be elicited at the earliest possible opportunity.

Figure 1: Phase IV TAMER

The third major component enabling remote operation was the pan-and-tilt unit, or PTU. The PTU provided the ability to change the orientation of the TAMER. For prototyping, a PTU unit from Directed Perception was chosen. The unit provided 340 degrees of rotation, and could tilt 60 degrees off the horizontal plane. The unit was rated to accommodate seven pounds, and was equipped with a RS-232 interface and a straightforward protocol command set. This command set incorporates pan, tilt, speed, and acceleration controls.

To implement the PTU mechanically, it was decided to build a PTU canister that would serve as both a PTU stand as well as PTU storage (see Figure 1). The canister was large enough so that when closed, it could hold the PTU unit (TAMER not attached), a BB-390A/U Nickel Metal Hydride rechargeable battery, necessary cabling, and a cell phone, all packed in foam. To deploy the TAMER, the canister top is removed by unscrewing three wing nuts and flipping the lid over so that the PTU is oriented upward. The wing nuts are then replaced, securing the lid. Three foldable legs on the bottom of the canister can be extended into a tripod configuration. The TAMER itself is attached to the vertical screw protruding from the PTU unit. The larger battery provides significantly more run-time than the BB516 batteries that normally power the TAMER.

For the TAMER to be used in a remote situation, an effective vision system is required. To implement this, we utilized the PCMCIA expansion capabilities and used a digital camera that interfaced through a PCMCIA card. The digital camera is capable of providing 312 X 287 pixel resolution, with 8 bits of radiometric quantization. The digital image is then transmitted to a central command station over a severely disadvantaged link (potentially as low as a 1,200 baud cellular connection). To transfer the 89,544 bytes of image data over a 1,200 baud connection would require approximately 10 minutes. To address the limited link bandwidth, an in-house wavelet image compression algorithm was developed. By compressing the image by a factor of 100, the image data to be transmitted is reduced to 900 bytes, which results in a 6-second transfer time. This image can then be decompressed at the receiver and displayed.

Distributed Interactive Simulation, or DIS, was another system interface aggressively pursued by the TAMER team. This effort was introduced because the TAMER project was ahead of schedule and under budget. Another reason was to better align TAMER with other business areas within Motorola. DIS is a network communications protocol that allows a network of devices to interact with each other to produce a battlefield simulation containing a large number of entities. A software interface was written to enable TAMER insertion into DIS exercises. This allows users to train with an actual TAMER without fielding the unit. It also allows the TAMER to participate in simulation exercises for demonstration purposes. A user can utilize the virtual DIS environment to provide the surroundings, while using the actual TAMER hardware. Commands (such as fire laser, acquire GPS, etc.) are intercepted by the DIS network, and realistic results corresponding to the TAMER's location and orientation within the DIS environment are fed back to the TAMER. Range and bearing information of virtual targets are obtained by simulating the interaction of the laser with targets in the virtual environment. DIS can be utilized to simulate the transmission of target position reports, as well as the transmission of images captured by TAMER's camera. The use of DIS also facilitates experimentation with enhancements such as the reception, transmission, and display of information, as well as other new capabilities.

Mechanically, the appearance of TAMER changed as new options were explored to improve the manufacturing process. An integrated chassis was developed to further reduce the number of separate parts that comprise the TAMER. In previous phases, TAMER was a module inserted between the battery module and laser module of the base-line MELIOS unit. The drawback of this method was that the team could not make modifications directly to the MELIOS, which, for example, forced previous phases to locate the control buttons on the bottom of the TAMER. Phase IV involved removing the electronics from the MELIOS battery module, and installing them into the newly designed TAMER integrated chassis. The chassis consisted of the main unit connected to the battery module. This reconfiguration resulted in substantially increased surface area. Accordingly, the VersaPoint user input module was located on the topside of the TAMER, over the battery compartment. This integrated chassis featured a flexible printed wiring board (PWB) card holder, and eliminated the need to screw individual PWBs to the cardholder. This card holder could slide into the main chassis when the laser module was not connected, thus simplifying the assembly of the TAMER system.

A number of other options were explored during Phase IV. The first of these options was to change the mode of operation from 5V to 3.3V. Most of the parts required to operate at 3.3V were readily available. The only part that could not be changed to the lower voltage was the floating-point unit (FPU) [11]. The FPU was added to decrease image compression execution time. Unfortunately, a suitable 3.3V FPU did not exist, and one would not be available until late in our phase IV development cycle. Accordingly, a 5V FPU was retained.

Another area of exploration was differential GPS (DGPS). The internal Oncore GPS of the TAMER is fully capable of supporting DGPS. In fact, the team successfully demonstrated DGPS capabilities by using two Oncore GPS units. One served as the base, and one served as the remote site. As a result, precision increased dramatically. A deployment scenario was created to supply DGPS to the battlefield using multiple TAMER units. One TAMER could be placed at a known position, and remote TAMER systems could dial into the base TAMER (via a wireless cellular link) and acquire the differential signals. In this way, DGPS could be supported on the battlefield with moderate additional complexity.

3. Methodology

The TAMER approach is a rapid and iterative development cycle. This methodology enables operational equipment to be generated quickly such that user feedback can be obtained. This feedback allows the TAMER team to start the next iteration with minimal delay.

3.1 Rapid and Iterative Development Cycle
The TAMER approach focused on maintaining frequent user interaction during the development cycle. The best feedback comes from giving the user a piece of equipment and allowing him to interact with it in his own environment. In the case of a dismounted soldier, this environment is the battlefield. To obtain feedback relevant to this environment, it was

required that the equipment was built in the final form factor. Final form factor allowed actual missions to be performed and analysis of the equipment to be supplied.

Any environment that inserts the user into the development cycle can expect frequent requirements changes, especially when no single person knows exactly what is needed. Engineers and users maintain open communication as new ideas are explored. Therefore, time is always a critical factor, and strategies need to be in place to minimize the amount of time required to reach the next stage of development. A critical balance among quality, performance, and functionality needs to be reached. For example, too much time spent on releasing a high-quality initial product could be wasted if the user does not like the new functionality. Conversely, too little time spent on quality can drastically increase the integration and testing time required to prepare the unit for user trials.

3.2 Hardware Methodology
Rapid prototyping methodology and the desire to freeze a final form factor (for accurate user feedback) drove engineers directly to the final PWB, with dimensions as defined in agreement with the mechanical engineers. Schematics were rigorously peer-reviewed to assist in compensating for the minimal use of breadboarding. Even so, it was not uncommon for multiple white wires to be added to the final board once actual testing was performed. In this type of development environment which can change rapidly based on user feedback, this was an expeditious and cost-effective method.

3.3 Software Methodology
The software methodology was based on the original Phase I TAMER code. That version of code set the base-line operating conditions of the system, and froze the general operating system to be used in future variants. To rapidly introduce new features at the earliest possible moment, a feature-focused development methodology was applied. Many of the features in the TAMER can run independently of each other. Because of this, focus can be placed on individual features, with less regard for interactions between features. While this produces a less robust system overall, it enables the introduction of new features quickly, which satisfies the goal of rapid development in a demonstration-oriented environment. This methodology has helped the software team develop a functional prototype such that user feedback could be obtained. This feedback can then be re-introduced into the cycle. If the feature was poorly received, then development can stop immediately. If the feature was perceived to be beneficial if modified in some way, then that feedback is put into the cycle. The next iteration can improve upon both the quality of the first iteration, as well as implement the desired modifications. If the feedback indicates that the feature is well received, the next iteration can improve the quality of the feature.

4. Power Conservation Strategies

4.1 Hardware
The process of power conservation began at the hardware level. A constraint set forth early in the project was that all TAMER power was to come from the same BB-516 24V NiCd battery that was used by the MELIOS. One reason for this was to demonstrate that TAMER was a viable addition to the base MELIOS unit. Another reason was the desire to use the existing batteries. Consequently, increasing the storage capacity of the system beyond the BB-516 24V NiCd battery was not an option. Therefore, to minimize the number of battery packs carried by the dismounted soldier, an aggressive power conservation program was initiated.

To that end, significant time was spent searching for low power components. Once these components were identified, analyses were performed to determine proper placement of these parts on the printed wiring boards such that power consumption was minimized.

Designation of I/O signals from the 68360 microcontroller was a vital part of the power conversation program. These I/O signals were used to turn off certain components when they were not needed. To achieve maximum power flexibility for the controlling software, intergroup coordination between hardware and software team members was a vital element during design. Both software and hardware team members had significant cross-competency. Blending these two disciplines proved to be the key ingredient in successfully designing the hardware to achieve a high degree of power savings. Without team members successfully working within their roles, the power savings to the degree experienced during this project could not have been obtained.

4.2 Software

The flexibility of the TAMER hardware platform offers a number of options available to the software designers to aid in system power management. These options include peripheral power management, system power management, and software optimization.

The first software solution addressed was peripheral power management. Peripherals are defined as anything external to the CPU. Peripherals would include items such as the GPS, the PCMCIA cards, the PCMCIA slots themselves, RS-232 transceivers, and the pan-and-tilt unit. For each device, different power modes were identified and analyzed. Time and effort for the transition between each of these states were also analyzed. Utilizing this data, power consumption and functionality were compared. Based on this comparison, the software was designed to effectively utilize each of the available states in order to maximize effective device utilization, while minimizing overall power consumption. To determine whether any signals could give the software additional flexibility, hardware design was also considered.

The second software solution addressed system power management. This area deals with issues relating to the CPU. The MC68360 offers a great deal of flexibility which gives the software a large number of power and performance options. One of the most important of these features is the ability to throttle the clock frequency. This can be done in one of two ways. The first method involves changing the phase-lock loop (PLL). The PLL multiplies an incoming clock frequency to a desired level. In this case, a low-cost 32.768KHz crystal provides the input, and the PLL can be given a multiplication factor that will produce a resulting clock speed between 10 and 25 MHz. The second method changes the system clock speed through the use of a clock divider. This divider works on the output of the PLL. The divider is a much cleaner method of slowing the processor, and provides less of an impact to other parts of the system. Additionally, the system can be put into an LPSTOP mode, which disables the PLL. This will essentially stop the generation of most system clocks until an interrupt is received. This mode is useful in times when the processor is otherwise idle.

The third software solution dealt with software optimization. With the large power benefits of having the system in an LPSTOP state, it is desirable to remain in this state as long as possible. For example, a full-screen write to the TAMER AMEL microdisplay (640 X 480) results in 307,200 iterations through a screen write routine. Using a standard nested "for" loop in C resulted in nearly 7,000 4K cycle-count blocks being executed. Using the "register" keyword in declaring index variables reduced the number of cycles to under 5,000 4K cycle-count blocks. Having the routine decrement instead of increment through the screen index reduced the count to under 4,000 4K cycle-count blocks. Using a decrementing index reduced the 4K-cycle-count block to under 2,000. By simply rewriting this routine, the execution time was decreased by a factor of three. Accordingly, the system can return to the power saving LPSTOP state much faster than before.

5. Image Coder Description

A custom state-of-the-art image compression algorithm was developed for the TAMER system. This algorithm employs a number of advanced technologies. Figure 2 shows a block diagram of the image coder. The coder is a non-entropy-coded design utilizing wavelet decomposition and trellis-coded quantization (TCQ). The coder combines attributes from the wavelet/TCQ coders reported in [12] and [13].

The image to be encoded is decomposed into 22 subbands using a modified Mallat tree configuration. That is, the image is initially decomposed into 16 equal-sized subbands, with two additional levels of decomposition being applied to the lowest-frequency subband. Each subband is collected into a one-dimensional sequence and encoded using fixed-rate TCQ (FRTCQ) [14] designed for the memoryless Laplacian source. All subbands are normalized by subtracting their mean (only the sequence corresponding to the lowest-frequency subband is assumed to have non-zero mean) and dividing by their respective standard deviations.

The total side information to be transmitted consists of the mean of the lowest-frequency subband, and the standard deviations of all 22 subbands. These quantities are quantized with 16-bit uniform scalar quantizers using a total of 368 bits. The compression ratio is also transmitted and allotted 10 bits. Finally, the initial trellis state for each encoded sequence requires 2 bits (for a 4-state trellis). The total side information then consists of 422 bits per image.

Fixed-rate TCQ codebooks are designed in one-bit increments from 1 to 8 bits/sample. The training sequence consisted of 100,000 samples derived from a Laplacian pseudo random number generator. Codebook design uses a modified version of the generalized Lloyd algorithm for vector quantizer design.

Rate allocation is performed by using the iterative technique discussed in [15]. The algorithm allocates bits based on the rate-distortion performance of the various trellis-based quantizers, and the energy content of the wavelet coefficients. This rate allocation procedure allows precise bit rate specification, independent of the image to be coded.

This image coding system allowed remarkable compression capabilities with impressive image fidelity. However, the inherent computational intractability of the algorithm required the addition of the MC68882 floating point unit to the TAMER system. The image coding system was implemented in floating-point math, and ran inefficiently on the MC68360. Interfacing with the MC68882 allowed the compression time for a 320 X 240 image to be reduced to 20 seconds on the computationally disadvantaged TAMER hardware. This time compares to less than one second on a 90 Mhz Pentium system. It should be noted that very little effort was expended on optimization of floating point protocols.

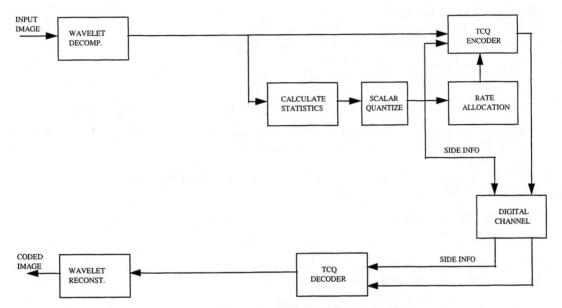

Figure 2: Wavelet/TCQ Image Coder

6. Conclusion

The objective of the TAMER program was to develop and demonstrate novel ways of combining sensors, computation, and communications in a lightweight, low power, modular package. The knowledge gained can subsequently be applied to other programs. For example, the controlling portion of the Lightweight Laser Designator Rangefinder (LLDR) is a follow-on to the TAMER project. The concepts of low power management, peripheral management, and domain knowledge are being applied to this portion of the LLDR design. Continued reuse of TAMER technology will contribute significantly to the utility of this and other programs.

7. Acknowledgements

This work was sponsored by the Advanced Research Projects Agency, Electronic Systems Technology Office under Contract MDA972-95-C-0012.

8. References

[1] Motorola, "Oncore User's Guide Revision 6.1," May 1994.

[2] Planar, "AMEL 640.480 monochrome Developer's Kit," Sep. 1997.

[3] G. Lutz, S. Hamilton, and L. Phillips, "TAMER Quick Reaction Demo Final Report," Aug. 1994.

[4] Motorola, "MC68360 Quad Integrated Communications Controller User's Manual," Mar. 1995.

[5] Collins Avionics & Communications Division Rockwell International Corporation, "PLGR Operations and Maintenance Manual, Satellite Signals Navigation Set AN/PSN-11," Sep. 1993.

[6] T.J. Jasa, "Interface Control Document, PLGR-to-PLGR & PLGR-to-PC interface," Jun. 1994.

[7] Motorola UDS, "CELLect 14.4 PCMCIA Installation and Operation," Mar 1994.

[8] Motorola, "Montana User's Guide," 1996.

[9] "PC Card Camera SDK 1.00 Programmers Guide," VLSI Vision LTD., Oct. 1994.

[10] "PC Card Camera 1.00 User Manual," VLSI Vision LTD., Oct. 1994.

[11] S. Harris and T. Johnson, "Software Links Math Chip to 68000-family Processors," Motorola, Inc., Jan. 1986.

[12] G. P. Abousleman, "Hyperspectral image coding using wavelet transforms and trellis coded quantization," *Wavelet Applications II*, H. Szu, Editor, Proc. SPIE 2491, pp. 1096-1106, 1995.

[13] P. Sriram and M. W. Marcellin, "Image coding using wavelet transforms and entropy-constrained trellis-coded quantization," *IEEE Trans. Image Processing*, vol. 4, pp 725-733, June, 1995.

[14] M. W. Marcellin and T. R. Fischer, "Trellis coded quantization of memoryless and Gauss-Markov sources," *IEEE Trans. Commun.*, vol. 38, pp. 82-93, Jan, 1990.

[15] Y. Shoham and A. Gersho, "Efficient bit allocation for an arbitrary set of quantizers," *IEEE Trans. Acoust., Speech, Signal Processing*, vol. 36, pp 1445-1453, Sept. 1988.

Recognition of human activities using handheld thermal systems

John D. O'Connor[a], Barbara L. O'Kane, Ph.D[b], Kathy L. Ayscue[a],
David E. Bonzo[a], and Beth M. Nystrom[a]

[a]E-OIR Measurements, Inc. Spotsylvania, VA 22553
[b]Night Vision Electronic Sensors Directorate, Ft. Belvoir, VA 22060

ABSTRACT

Development of low-cost lightweight thermal sensors necessitates the re-evaluation of detection, classification and recognition criteria as applied to human performance with thermal imaging systems. The need exists to assess spatial and motion characteristics of man targets, rather than vehicle targets, and how these differing characteristics affect human perception and performance models. Higher order discriminations, such as determining activities and objects being carried are useful in determining range performance of specific sensors, and reflect the needs of the man-portable sensor user.

Keywords: human target, detection, classification, recognition, identification, cycle criteria, handheld thermal sensors

1. BACKGROUND AND PURPOSE

Advances in thermal imaging technologies have made possible man-portable, durable, dependable thermal imagers. Large platforms, such as tanks, helicopters and aircraft, are no longer required to field these imaging systems. This shift from platform-based to man-portable capability brings about new criteria for target detection, recognition and identification. While platform-based systems are designed and tested with vehicle or aircraft targets as the primary target objective, man-portable systems must also enable users to accurately understand the nature of human activities to be effective on the battlefield.

A significant body of experimental data exists for the probability of detection, recognition and identification of vehicle and aircraft targets at various ranges and environmental conditions with different imaging systems, few studies have been done to provide experimental data for human targets[1,5]. Also, while a definition of recognition and identification are established for vehicles (recognition being functional category discrimination (Tank, Truck, Armored Personnel Carrier, identification being exact nomenclature (M1 vs M60 vs T62), even a definition of human target identification with thermal sensors has not been explored until this test. Detection, recognition and identification ranges can be predicted with the ACQUIRE model[6] for vehicle targets because these definitions exist, but definitions for the human target have been left at simply detecting a human in the field of view. Human target recognition and identification are inherently dissimilar to vehicle or aircraft recognition and identification. Human targets are much smaller, and are often characterized by lower mean temperature differences when compared to the background environment. Human targets are also more difficult to classify as friendly, neutral or hostile. While the type of vehicle or aircraft can indicate a target as friendly, neutral or hostile, a human target must be classified primarily by observed behaviors and secondarily by identifiable armament or equipment.

Many applications exist wherein the range at which an observer would be able to go beyond detection - to perceive more about the person in the field of view - would be a useful piece of prediction information. However, no definition had been established heretofore. Many possibilities existed for defining "recognition" or "identification" of a human. The most feasible suggestion for operationalizing this concept of human target "recognition" was suggested in 1993 by Loren Doyle. It used the definition that recognition occurs when the observer can perceive what the person is doing. This seemed like a good definition to test the handheld sensors against to see whether range performance curves would be well-behaved and meaningful with such a definition. If the definition was reasonable, subjects should find the task relatively natural to perform, the resulting "recognition" curves should be predictable with the expected shape of dropping off gradually with range, and the curves should be below the detection curves. Finally, it was hoped that a cycle criterion could be calculated which would fit the data in a reasonable way.

Part of the SPIE Conference on Sensor Technology for Soldier Systems • Orlando, Florida • April 1998
SPIE Vol. 3394 • 0277-786X/98/$10.00

51

2. IMAGERY COLLECTION AND EDITING

The imagery collection effort took place outside of Darwin, Australia approximately 1700 miles north of Adelaide along the northern coastline. This location was chosen due to the tropical nature of the foliage and the weather conditions. In this harsh tropical environment expected day and night temperatures were 32 and 26 degrees C, respectively, with a relative humidity averaging above 85%. The imagery used in this study was collected on March 27, 1995 from 0930 to 1430 hrs local time. The temperature was 32 degrees Celsius, and the relative humidity was between 80 and 95 percent. Visibility was good and clouds were scattered.

Engineers and scientists from NVESD, Ft. Belvoir, VA, employed two thermal imagers: a 3-5 micron thermoelectric cooled FLIR and an 8-12 micron uncooled FLIR. The 3-5 micron system was used only in the NFOV. Thermal imagery was captured on Super-VHS tape from outputs built into each sensor.

The scenarios used in this study involved between 1 and 3 people performing 14 different activities at each of nine ranges. Activities were performed by Australian enlisted personnel. These personnel were consistent in repetition of the scripted scenes. Movement directions, lengths and speeds were nearly identical at the differing ranges in each code-specific scene. The ranges employed will be discussed here only as ranges 1 through 9. The order of activities was randomized at each range. Each activity, or scene, was identified by the following code key found in Table 2.

Table 1: Activity Codes

Code	Activity
A1	Two people approach each other and pass a package between them
A2	A walking person carrying a stick
A4	A kneeling person pointing towards the sensor
A5	A walking person carrying a 1m section of drain-pipe
A7	Two people approach each other and shake hands
A8	Two people unloading two large boxes from a vehicle
A9	Two people walking
F2	Two people unloading a large box and a rifle from a vehicle
F3	A walking person with a shoulder-launched weapon slung on right shoulder
F4	A captor with rifle marching a prisoner
F7	A person moving in patrol order with a rifle
F9	Two people approach each other and pass a knife or pistol between them
F10	A kneeling person with rifle pointed towards the sensor
F11	A kneeling person employing a pair of binoculars or night vision goggles
X0	No people, activity, or items

* Activities coded with an "A" are considered friendly or neutral;

* Activities coded with an "F" are considered hostile or foe.

Each scene was edited to a viewing length of between seven and twelve seconds, with three to four seconds of blank tape between scenes, and six to eight seconds of blank tape between ranges. A background scene with no activity, people, or items was edited into each range series and coded as X0.

3. TEST METHOD

Since the subjects would be viewing a large number of scenarios, a quick response method was developed to reduce the amount of writing which would be required in the semi-darkened test room. Subjects were to choose among activities, as shown in Table 3. Subjects were expected to use the answer key to score each scene of the video imagery based on four criteria. The first criterion scored was the number of people visible in the scene. Subject choices were 0, 1, 2, 3 or?, where the question mark denotes an inability to distinguish the number of individuals. The second criterion was the visible activity in the scene. Subjects were expected to choose from a list lettered A through N in which all possible scene activities were included, plus two activities not in the imagery set. Choices of unsure and none were also included. The third criterion was the visible items used or carried in the scene. Subjects were to choose from a list of 10 possible items, eight of which were actually visible in some scenes. Choices of unsure and none were also included. The last criterion was the subject's rating of a scene as a military or non-military activity. Subjects would be asked to place a check mark or X in a box denoted "military activity" if they believed a weapon or sensor was present in the scene.

Table 2: Key of responses for answer sheet. Subjects were to choose the number of people, one or more activities and one or more objects carried for each trial.

# People	Activity A-N	Objects Carried 1-12
0,1,2,3,?	A. Aiming Weapon	1. Boxes, packages
	B. Crawling	2. Briefcase/ suitcase
	C. Kneeling	3. Knife/ pistol
	D. Passing item	4. Night vision goggles
	E. Patrolling	5. Pipe
	F. Pointing	6. Radio
	G. Prisoner march	7. Rifle
	H. Shaking hands	8. Shoulder launched weapon
	I. Standing still	9. Shovel
	J. Unloading items	10. Stick
	K. Walking	11. Unsure
	L. Waving	12. None
	M. Unsure	
	N. None	

Subject testing took place on 25, 26, and 27 July 1995 at Ft. Bragg in Fayetteville, North Carolina. The NVESD perception group asked for and received 60 test subjects from enlisted personnel of the 2nd Brigade, 82nd Airborne Division stationed at Ft. Bragg. 30 subjects had some prior experience with thermal imagery and 30 had no such prior experience, as requested. Subjects experienced and inexperienced in thermal imagery were randomly mixed in the test groups. Testing was scheduled in 4 sessions of 3 hours, with a different group of test subjects in each session. The first session, with 13 subjects, was a pre-test, with the goals of identifying any deviations from the original testing plan, and ensuring the proper working order of the video equipment. Due to some equipment malfunctions with the TV monitors, data collected from these subjects was not incorporated into the data base. The following three sessions involved 47 subjects, in groups of 15, 15 and 17.

Testing was conducted in an air-conditioned room with windows on the north side. Four nineteen-inch Goldstar television monitors received video input of the thermal imagery from a single Sony VCR, so that all subjects being tested at the same time saw the same imagery simultaneously. Four or five chairs were placed in an arc equidistant from each monitor at a distance of 1.5 meters. All monitors were placed on tables of approximately 80 centimeters in height, covered by black cloth. One monitor was located near the east wall facing west, one in the room center facing west, and two near the south wall facing north. Window blinds were adjusted so that no glare was apparent on any monitor. VCR tape heads were cleaned prior to testing. All monitors were adjusted to the same brightness, contrast and tint levels.

Subjects signed an attendance sheet upon arrival and were given a subject number. They were then informed that the purpose of the test was to determine the ranges at which soldiers could accurately determine the number of human targets, the activities, and the items visible or carried in a thermal image scene, and whether an activity depicted was of a military nature, (defined as involving a weapon or sensor). Each subject was asked to score two-hundred-sixty-five scenes to the best of their ability within the given criteria. Subjects were allowed to move as close to the monitor as they wanted to, as long as they did not block the view of any other subject. Subjects chose their own seats, as long as no less than three and no more than five subjects viewed any one monitor. Subjects were asked not to discuss their perceptions or look at any other subject's answers. Subjects were told that their test performance would not have any bearing on performance ratings or any other type of formal or informal professional evaluation, but that any disruptive or uncooperative behavior would be reported to the base commander. All subjects were cooperative and followed instructions. Subjects were asked to rate each scene according to the four test criteria outlined on their answer keys. Subjects were told to circle more than one answer in the activity or object column if they perceived more than one. They were asked not to guess and to mark the unsure answer when they were unsure. Subjects were informed that because different ranges would be viewed, activities at some ranges should be more difficult to discern. Subjects were informed that four sets of imagery with sixty to seventy-five scenes in each set would be viewed over a three hour period, and breaks of ten minutes would be taken between imagery sets. Subjects were told that they would have fifteen seconds to mark their answer sheet after each scene and that each scene would be viewed only once. Subjects were instructed that if they did not, for any reason, view a full scene they should mark no answers for that scene and indicate this by marking through that scene number. Subjects were allowed to ask questions about the procedure and purpose of the test before testing began. A sample answer was demonstrated before testing began, and each individual's response was checked to verify understanding of test instructions.

Each subject group viewed the imagery sets in a different sequence. The first viewed in the order ABCD, second in the order BCDA, third in the order CDAB and the last in the order DABC. Subjects were monitored by one NVESD researcher while another controlled the VCR and monitored image quality. If an image was distorted or imperfect, subjects were instructed to skip that scene on their answer sheet and continue to the next scene. One subject did not follow instructions properly, so this data was discarded as unusable. One subject arrived as testing began. He was allowed to view the first image set, fully instructed during the break, and scored scenes in the second, third and fourth sets of imagery. This data was included where complimentary ranges existed for each sensor. Each group completed testing within the expected time period. Completed answer keys and questionnaires were collected and labeled for each individual subject, sorted by subject group, and returned to Ft. Belvoir for analysis.

4. RANGE PERFORMANCE OF THE SENSORS

The probability of detecting the correct number of people, the correct action(s) of the people, and the correct object(s) carried were calculated for each scenario. An overall average of all scenarios at each range was calculated and the results plotted in Figures 1a and 1b for the 3-5 micron thermoelectric cooled and an 8-12 micron uncooled systems respectively.

The expected ranges for 50% probability of detection are 9 and 7, respectively, for the 3-5 micron and 8-12 micron systems under ideal environmental conditions. In this study the 50% probability ranges for detection were 9 (extrapolated) and 8, respectively. The agreement between expected and obtained values is gratifying because the imagery used in this study was from a more difficult scenario: relatively high clutter, high temperature and humidity and in the daytime, far less than ideal conditions for optimal thermal sensor performance. The thermal signature of a human target in such an environment has a far lower mean temperature difference from the background, and the high humidity levels inhibit infra-red transmission significantly. Also, noted previously[1], these sensors do only about half as well during the day as at night.

Figure 1a: 3-5 Micron Thermoelectric Cooled Overall Scenario Performance Average

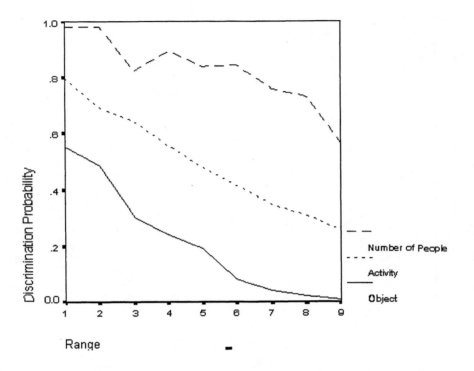

Figure 1b: 8-12 Micron Uncooled Overall Scenario Performance Average

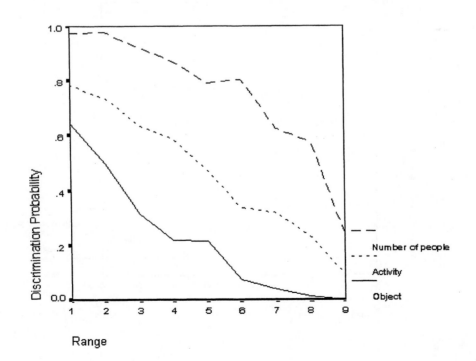

Figure 2a: 3-5 Micron Thermoelectric Cooled Walking Scenarios Performance Average

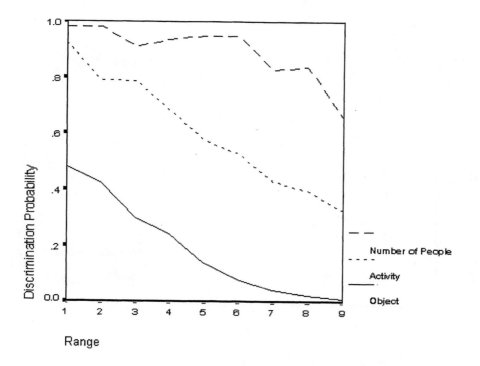

Figure 2b: 8-12 Micron Uncooled Walking Scenarios Performance Average

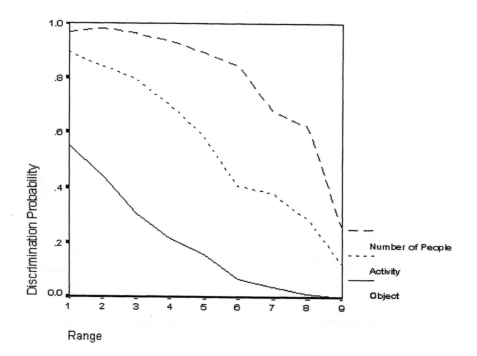

The type of activity in each thermal scene greatly influenced the subjects' ability to discriminate the number of individuals, the activities and the items present. Scenes in which the primary activity is walking had the overall best average detection and recognition of the activity, especially at the more distant ranges when performance against other types of scenarios was much reduced. This was especially true for the 3-5 micron system which seemed to have greater range performance for detecting the number of people in the scene for walking scenarios. The 8-12 micron system appeared to drop off more steeply and at a closer range. Figures 2a and 2b show the range performance results for all the scenarios involving people walking and performing various activities. It should be noted here that this data is not presented with the assertion that universally 3-5 micron systems outperform 8-12 micron systems for this task. In this particular study this particular 3-5 micron system outperformed this particular 8-12 micron system with this particular task under these particular conditions. It is not the authors' intention to assert superiority of 3-5 micron over 8-12 micron systems, or cooled over uncooled systems.

This high level of detection for the walking personnel scenes can be explained by the simple premise that an object with a distinct thermal signature moving across a random background would be easier to locate than a static one. The long range recognition of walking as a primary activity can be accounted for by the familiarity of the subjects with the unique motion characteristics associated with the bipedal locomotion of a human. This is borne out by the fact that, for scenes in which walking was intrinsic to the activities, detection range stayed high over long ranges, while recognition of the primary activity did not.

In contrast, scenes in which the primary activity involved kneeling had the lowest average detection and recognition. This trend became more pronounced as range increased. (See Figures 3a and 3b.) There are many reasons why this target would not be detected well. First, obviously, a kneeling target is harder to detect because of the smaller area exposed in the field of view. Second, the lack of motion makes discrimination from the background more difficult. Third, a kneeling target loses much of its apparent body symmetry and expected aspect ratio, in effect becoming more random in shape and thus more similar to a natural background object.

The vehicle unloading scenes are of special interest. In these scenes, one person stands at the back of a truck and hands two items to another who walks a short distance and places them on the ground. The truck is not in the field of view at range 1 for either sensor, and this would explain the low recognition scores for these activities at range 1. Subjects could only infer the presence of the truck, and thus often scored this scene improperly. At range 4 and beyond, subjects had difficulty detecting the correct number of individuals in the scene, even though they recognized the activity. It is possible that the individual at the back of the truck blends in with the truck and is thus harder to distinguish from it.

Identification of the specific items appears highly range limited and shape dependent. The large boxes, (approximately 0.5m x 0.5m x 0.5m), of the unloading scenes were the only items which had a probability of identification of 50 percent or greater beyond range 4. This is likely due to their size and cuboid shape. It is important to note that the small package, (about 12cm x 12cm x 12cm) employed in the package passing activity was more easily identified at longer ranges than items like the rifle or shoulder launched weapon, which are many times larger, or the knife or pistol, which are comparable in size. The small package may be more easily identified because its sharp edges delineate it from its background. Scenes involving the carrying of a rifle in patrol order had significantly higher item identification than scenes in which the rifle pointed towards the sensor, or was unloaded from the truck. This is partially dependent on the aspect of the rifle, but may in large part be due to the way the rifle was being carried. The shoulder launched weapon was of similar size and was displayed in similar aspect, but was much harder for the subjects to identify. Some of this discrepancy may be due to the fact that this weapon was shoulder-slung. Had the weapon been carried in patrol order, it may have been easier to identify.

Figure 3a: 3-5 Micron Thermoelectric Cooled Kneeling Scenarios Performance Average

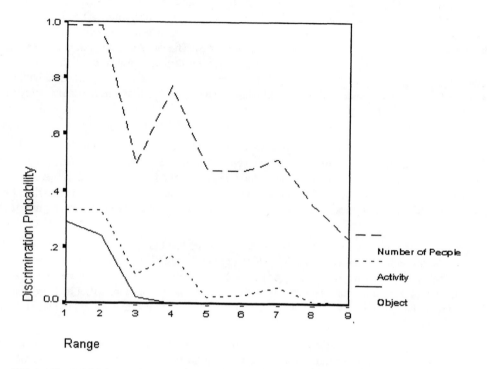

Range

Figure 3b: 8-12 Micron Uncooled Kneeling Scenarios Performance Average

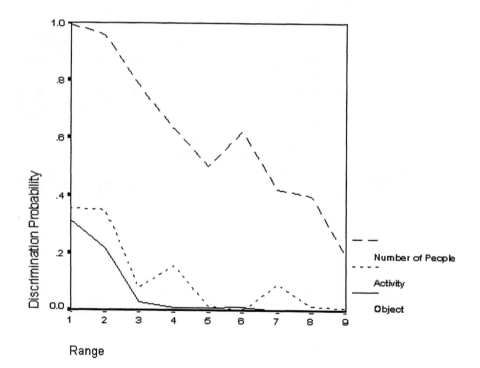

Range

5. MODELING

To understand the perception of humans through a thermal device for the purpose of determining intent or activity was one of the purposes of this experiment. The hope was that modeling parameters could be developed which would apply to this perception and that therefore predictions could readily be made as to the ranges at which a person's activity including what he was carrying could be discerned. Prior to running the experiment, the ACQUIRE model was run using a hypothesized set of criteria. In this section, we will describe the traditional ACQUIRE model methodology and compare predictions based upon various cycle criteria to the obtained results from the subjects in the study.

The Night Vision target acquisition model for sensor range performance, ACQUIRE, predicts detection ranges based on the well-known Johnson criteria, where the probability of detection is determined by the size of the target, its mean temperature difference from the background and the range to the target. A probability function, known as the Target Transform Probability Function, can then be predicted. The cycles on target, N, is calculated as the product of the resolvable frequency from the 2-dimensional MRTD curve and the square root of the target projected area divided by the range (km). N_{50} is the number of resolvable cycles on target required for 50% of an ensemble of trained observers to be able to perform the task of detection, recognition, or identification. Standard values for N50 are 0.75, 3.0 and 6.0 for detection, recognition, and identification, respectively, for an ensemble of targets in low to medium clutter.

a. Detection -- determination that a target of military interest is present

b. Recognition -- determination that a target belongs to a particular functional category (for example, truck, tank or armored personnel carrier)

c. Identification -- correct judgment of the exact nomenclature of the vehicle (T62, M35, M60, etc.)

Each task is assigned an empirically derived standard cycle criterion (N_{50}) as shown in Table 3

Table 3: Recommended Cycle Criteria (N_{50}) for static discrimination tasks.

Task	Cycle Criterion
Detection	0.75
Recognition	3.00
Identification	6.00

The criterion number of cycles on target required for the various tasks of detection, recognition, and identification have been the subject of experimentation which will be described further on in this report. The recommendations for the present ACQUIRE model for detection are based on an ensemble of tactical vehicle targets for a static discrimination task.. The interaction of factors such as type of task, observer expectation, motion, clutter level, amount of search and objects within the scene will affect detection probabilities.

Under this modeling algorithm, the criterion cycles on target (N_{50}) for detecting a human has been found empirically to be between .35 and .55 cycles on target[5] regardless of whether the human was moving or standing still. When the contrast was lower, moving targets however, were more easily detected than still targets[5]. The standard parameters used in modeling a person are .75 m^2 for the characteristic dimension, and 2 degrees Celsius to ensure that the model is not overly optimistic for military scenarios. The difference in modeling between the kneeling and standing scenarios is that the characteristic dimension for the standing man is .75 m^2 and the kneeling man is .5 m^2.

The question that is raised continually concerning the perception of humans is how to define the tasks associated with the perception so that useful modeling predictions can be made. The hierarchy for human target range performance prediction modeling is a departure from vehicle classification doctrine because of the inherent differences between human and vehicle targets. The criteria that apply to vehicle discrimination are inadequate when applied to human targets, and thus human target discrimination must be founded on new criteria that are based on detectable activities and items. For example, it was hypothesized that instead of .5 - 1.0 cycles on target as for detection, detecting the activity of the person (call it "recognition") might require two cycles on the target. In a first attempt to apply the model to these results, the standard modeling predictions were used. These predictions were somewhat in agreement with the results for the two sensors but significant variance remains to be explained.

Detection of the Number of Personnel in the Field of View: Both the upright and non-kneeling target scenarios varied somewhat from the predictions. For example, in general, especially for the upright targets and the 3-5micron sensor, the drop off in performance with range is more gradual than that predicted by the model. However, this agrees with previous findings that moving targets are more easily seen as the contrast drops off (with range, for example). This does not explain why the range performance with the 8-12 micron imagery shows a more rapid decline with range in this tropical daytime environment than the 3-5 micron imagery. For kneeling scenarios, a slightly higher cycle criterion seems to be required. Nevertheless, if all cycle criteria are fit to the data, the modeling results agree best with the traditional modeling parameters of .75 cycles on target for detection of upright, moving people, and about 1.0 cycle for the kneeling, non-moving targets. However, as stated above, the fit is by no means perfect, and more modeling enhancements would be needed. Unfortunately, the field data did not include standing people who were not moving, which would have yielded very important comparisons for modeling.

The ranges approximate the required specs for the systems and are twice as good as those found in prior studies with much earlier versions of these systems[1]. Also, in the earlier study, it was found that the range for detection was twice as long at night as during the day. This suggests that the performance should be even better at night than found here since the thermal contrast should be considerably greater.

Recognition of Activity: In this study we operationally defined "recognition" of a person as being able to discern the activity. This seemed to work quite well in that the observers were able to discern activities of the "target personnel" in a fairly reliable fashion as a function of range, with approximately 2 cycles on target being a fair approximation of the cycle criterion for upright subjects. The recognition function followed approximately the same overall shape as the detection function with a substantial reduction in performance at the middle ranges of performance.

Identification of Object Carried: At ranges of 1 and 2, the best that can be expected is a probability of 0.65 and 0.55 for the 8-12 micron and 3-5 micron systems, respectively, with both reaching a probability of 0.10 at range 4. The 50% probability was range 2 for both sensors. Some of the scenarios were relatively easier: passing a package, a captor with rifle marching a prisoner, persons moving in patrol order with rifle, and people unloading large boxes from a vehicle. In other scenarios the objects were much more difficult to identify: a stick, shoulder launched weapon on right shoulder, rifle pointing toward the sensor, binoculars, or a knife or pistol. As a general rule, the larger and more familiar an object, the more likely it is to be identified, especially if the context of the activity (e.g., patrol march) would suggest it. The smaller, less familiar and less expected an object is, the less likely it would be identified. What we do learn from this is that only at a very short range may many objects of military interest be identified. Because of the extreme variations between objects, cycle criteria for object identification will not be elaborated further in this document.

6. SUMMARY

In summary, the critical criteria for human target detection, as defined here, are target motion, posture, and range. For recognition, familiarity of the observer with the activity being observed also seems to be very important to ability to recognize the activity at a given range. While item discrimination is dependent on the shape and size of the object, the range, and its aspect relative to the observer, it is also apparently related to how the visible item is being used or carried. The sensor systems studied here are designed for use as a rifle-mounted or man-portable system. In the field, detection alone is not sufficient justification to fire if friendly or civilian casualties are to be avoided. There are a great many more modeling issues which are raised by this paper. However, these will be addressed by the overall modeling program. The database of man target imagery and results is now available for the modeling community and this report documents the details of the experiment for use by the community. Such issues revolve around the appropriate shape of the function for target detection, effects of posture, range, and target motion.

7. ACKNOWLEDGEMENTS

This project was sponsored by the Project Manager, Night Vision/Reconnaissance, Surveillance and Target Acquisition. The authors gratefully acknowledge their support.

8. REFERENCES

1. J. Farr, "Final Report: Short Range Thermal Sight (SRTS) Characterization." Technical Report NV 1-4, Mar 1987, Night Vision Laboratory, Ft. Belvoir, VA.
2. R. Hudson, *Infrared Systems Engineering.* c. 1969 John Wiley and Sons New York, Chichester, Brisbane, Toronto, Singapore.
3. J.Johnson, *Analysis of Image Forming Systems.* Ft. Belvoir, VA 1958.
4. J. Johnson and W. Lawson, *Performance Modeling Methods and Problems.* ISIS Specialty Group on Imaging, 1974.
5. B. O'Kane, M. Crenshaw, J. D'Agostino and D. Tomkinson, "Human Target Detection Using Thermal Systems." Ft. Belvoir, VA 1991.
6. J. Ratches, W. Lawson, L. Obert, R. Bergemann, T. Cassidy and J. Swenson, "Night Vision Laboratory Static Performance Model for Thermal Viewing Systems." Ft. Monmouth, NJ 1975.
7. F. Rosell, "Levels at Visual Discrimination for Real Scene Objects vs. Bar Pattern Resolution for Aperture Noise Limited Imagery." Ft. Belvoir, VA 1975.
8. F. Rosell, "Performance Synthesis of Electro-Optical Sensors." Ft. Belvoir, VA 1977.
9. F. Rosell, "Static Field Performance Models." NRL Report 8311, Washington, DC 1978.
10. C. Royal, "OCONUS Trip Report." Ft. Belvoir, VA 1995.
11. L. Scott, "Laboratory Evaluation Team Test for Thermal Imaging Systems." Ft. Belvoir, VA 1994.
12. G. Waldman and J. Wooten, J. "Electro-optical Systems Performance Modeling." c. 1993 Artech House, Inc. Norwood, MA.
13. W. Wolfe and G. Zissis, *The Infrared Handbook.* Office of Naval Research, Department of the Navy, Washington DC 1985.
14. ACQUIRE Range Performance Model, Version 1. U.S. Army CECOM RDEC NVESD Ft. Belvoir, VA 1994.

The Integrated Sight: Future Improvement for Land Warrior

Joe S. Randello and Lawrence T. Marshall
CECOM NVESD, Fort Belvoir, Virginia

Mario E. Velez
PM Soldier, Fort Belvoir, Virginia

Bob Frink
Raytheon TI Systems, Inc., Plano, Texas

Abstract

The Integrated Sight (IS) is a Technology Demonstration program within the Force XXI Land Warrior Program. Like the other Force XXI Land Warrior components, the IS is a candidate for future technology insertion into the Land Warrior (LW) program. The IS integrates an uncooled thermal imager, eye-safe laser rangefinder, electronic compass, CCD camera and infrared laser pointer into a single lightweight sight. It can mount on weapons or be used in a handheld or tripod mode. The functionality of the IS provides the fighting soldier with the ability to acquire targets during daylight, darkness, adverse weather, and through battlefield obscurants to provide target position data (azimuth and range) for indirect fire. Standardized interfaces enable digital target location data and imagery to be transmitted to and from the LW computer/radio subsystem (CRS). By integrating the sensors into a single unit, the IS offers reductions in weight, power, and size which is expected to improve the LW soldier's fightability. The impact of sensor integration on the LW soldier's tasks will be evaluated during Force XXI Land Warrior testing. The IS is being developed jointly by Raytheon Texas Instruments Systems (RTIS), the US Army CECOM RDEC Night Vision and Electronic Sensors Directorate, and the US Army Soldier Systems Command. This paper provides a comparative assessment of the IS and the LW baseline Weapon Subsystem, and outlines the expected improvements that the IS provides the fighting soldier.

62

Part of the SPIE Conference on Sensor Technology for Soldier Systems ● Orlando, Florida ● April 1998
SPIE Vol. 3394 ● 0277-786X/98/$10.00

INTRODUCTION

The Integrated Sight (IS) is a Technology Demonstration (TD) program being developed under the Force XXI Land Warrior (LW) program. The program is developing hardware that has been identified as a technology insertion candidate to the LW program as a potential upgrade of the LW weapon subsystem components.

Land Warrior is the Army's program to develop and field a totally integrated soldier fighting system by the year 2000. The LW system integrates commercial, off-the-shelf technologies into a complete soldier system. The LW system is comprised of five subsystems: the Integrated Helmet Assembly subsystem (IHAS), the software subsystem, the Computer/Radio subsystem (CRS), the Protective Clothing and Individual Equipment subsystem (PCIE), and the weapon subsystem. The IHAS provides a light ballistic protection helmet shell, and a helmet mounted display for interfacing the soldier to the other subsystems and to the digital battlefield. The software subsystem controls the soldier's core battlefield functions, display management, and mission data and equipment. The CRS integrates all of the electronic portions of the LW system. The CRS consists of a squad and soldier radio, a Global Positioning System (GPS), a handheld flat panel display and a computer. The PCIE provides the mechanical means of integrating the LW system. The weapon subsystem consists of a thermal imager, a CCD Camera, a close combat optic, an eyesafe laser rangefinder/digital compass assembly (LRF/DCA), and an infrared laser pointer. The weapon subsystem components are mounted on an M16 or M4 based Modular Weapon System.

The LW weapon subsystem provides the soldier with enhanced detection and recognition capability of man and vehicle size targets over current image intensifier sights in different types of terrain, vegetation, climatic and weather conditions. When used in conjunction with the soldier's Global Positioning System (GPS) and the CRS, targeting information for indirect fire from artillery, mortars, and aircraft can be acquired. The thermal imager is used primarily for night and limited visibility operations, the CCD Camera and close combat optic for day operations, the LRF for precise target range determination, the digital compass for target azimuth and elevation determination, and the laser pointer for inter-squad target marking.

The intent of the IS program is to allow the Army to evaluate the concept of integrating the multiple sensors of the LW weapon subsystem into a single weaponsight. The IS currently combines a thermal imager, a CCD Camera, an eyesafe LRF, an electronic compass, and a laser pointer. The IS provides the soldier with the same capability as the LW weapon subsystem, but offers reduced weight, size and power. Considering that weight, size and power are three of the highest priority issues for soldier equipment, the IS deserves serious consideration for future LW systems.

The focus of this report will be on the potential improvements the IS, as an integrated system, may provide the soldier in the future compared to the LW weapon subsystem's modular approach. Note: this paper represents a point in time of evolving hardware and should be read as an exploration of concepts and areas for evaluation, not a definitive comparison of the two configurations.

DESCRIPTION OF THE IS AND LW WS

The thermal imager allows the soldier to detect and recognize targets during the day, night, and through battlefield obscurants such as smoke and dust. The IS thermal imager uses an uncooled staring focal plane array, operating nominally in the 8 to 12 micron spectral region. The objective lens assembly collects infrared energy and focuses it through a chopper assembly onto the detector assembly. The chopper is used to remove the average thermal background component of the scene, resulting in the target temperature difference being measured. The focal plane array size is 245 by 328 square pixels. LW uses the Thermal Weapon Sight (TWS) as the thermal imager. The TWS uses a second-generation photovoltaic Mercury Cadmium Telluride focal plane array that is thermoelectrically cooled and operates in the 3.4 to 4.2 micron spectral region. The TWS is a serial scan device and has an array size of 40 by 16 square pixels. The IS uses a cathode ray tube and the TWS uses light emitting diodes for the display. Performance requirements for the IS and LW are based on typical individual served weapon engagement ranges. Table 1 lists characteristics of the IS and the LW thermal imager (Medium TWS).

Parameter	IS Performance	LW Performance
Recognition Range	70% Confidence	70% Confidence
Man	800m	1100m
Vehicle	1000m	1100m
FOV		
Narrow	9° (H) x 6.7° (V)	9° (H) x 5.3° (V)
Wide	None (Single FOV)	15° (H) x 8.9° (V)
Resolution	0.5 mrad	0.6 mrad
NETD	0.15°C	0.1°C
Spectral Region	8 to 12 um	3.4 to 4.2 um
'Standby' to 'On' time	4.5 seconds	3 seconds

Table 1: IS and LW TI Characteristics

The CCD camera allows the soldier to detect and recognize targets during the daytime. The IS CCD camera uses an off-the-shelf COHU 1100 series CCD camera. The IS and LW CCD camera imaging sensor both use a 1/3 inch format interline transfer CCD with microlens-enhanced sensitivity. The resolution for both is 580 horizontal by 380 vertical television lines. The CCD camera is sensitive to the infrared laser pointer energy and the IS boresight procedure exploits that capability to simplify the IS boresighting procedure. Table 2 lists characteristics of the IS and LW CCD camera.

Parameter	IS Performance	LW Performance
Recognition Range Man Vehicle	50% Confidence 400m 1900m	50% Confidence 360m 1700m
FOV	9.0° (H) x 6.7° (V)	10° (H) x 7.5° (V)
Resolution	580 (H) x 380 (V)	580 (H) x 380 (V)
Pixels	768 (H) x 494 (V)	768 (H) x 494 (V)
Shutter	Electronic IRIS	Electronic IRIS
AGC	20 dB	20 dB
Magnification	3x	3x

Table 2: IS and LW CCD Camera Characteristics

The laser rangefinder is used to accurately determine the range from the soldier to targets of interest. The IS LRF design incorporates direct diode pumping of a Neodymium: Yttrium Lanthanum Fluoride (Nd: YLF) laser rod. The laser energy is shifted from 1.047 microns to the 1.54 micron wavelength using an intercavity Potassium Titanyl Phosphate optical parametric oscillator. The LW LRF is also a diode pumped, passively Q-switched laser but the lasing medium is erbium. Consequently, there is no intermediate frequency; the erbium lases at 1.537 microns. The output wavelength and output power of both the IS and LW LRF have been chosen to ensure that the designs are eyesafe while meeting the ranging requirements. Table 3 lists characteristics of the IS and LW LRF.

Parameter	IS Performance	LW Performance
Maximum Range	25 m to 2500m	25 m to 2500m
Resolution	± 1 m	± 3 m
Output Power	0.5 MJ	0.36 MJ
Wavelength	1.54 um	1.537 um
Eyesafety	ANSI Class I Eyesafe	ANSI Class I Eyesafe
Firing Rate	10 shots per minute	10 shots per minute

Table 3: IS and LW LRF Characteristics

The electronic compass determines the bearing to a target of interest. The IS electronic compass uses three orthogonally oriented magnetic sensors and three orthogonally oriented piezoresistive accelerometers to determine the azimuth, vertical angle and cant. The LW digital compass uses three orthogonally oriented magnetic sensors and two orthogonally oriented piezoresistive accelerometers. In order to meet the azimuth accuracy, both systems must compensate for both "hard" and "soft" iron effects, for the magnetic influences of other subsystems, and for the effects of the local area. Compensation is achieved using complex algorithms and a calibration procedure. Table 4 lists characteristics of the IS and LW electronic compass.

Parameter	IS Performance	LW Performance
Azimuth accuracy	±10 mils (±0.6 deg)	±10 mils (±0.6 deg)
Vertical angle accuracy	±3 mils (±0.2 deg)	±10 mils (±0.6 deg)
Vertical angle range	±90 deg	±45 deg
Data Rate	60 hz max	60 hz max
Settling Time	0.016 sec	0.5 sec

Table 4: IS and LW Electronic Compass Characteristics

The infrared laser pointer enables the operator to mark targets at night with energy that is visible to soldiers using image intensifiers. Those soldiers can then direct coordinated fire to the same target. The IS and the LW laser pointer are a direct diode based design and output power is limited to maintain eyesafety. The LW infrared laser pointer is the fielded AN/PAQ-4C. The IS uses the same electronics, optics, and boresight features as the AN/PAQ-4C laser pointer but uses a different packaging and mounting approach. Table 5 lists characteristics of the IS and LW infrared laser pointer.

Parameter	IS Performance	LW Performance
Output Power	1 mW	1 mW
Wavelength	840 nm	840 nm
Modulation	CW	CW
Eyesafety	ANSI Class I Eyesafe	ANSI Class I Eyesafe

Table 5: IS and LW Laser Pointer Characteristics

WEIGHT

Soldier borne weight is a critical issue in the design of new soldier equipment. The advent of new technology has provided soldiers with more capability, but at the same time caused an increase in the carrying load. The US Army Infantry Center has identified the reduction of soldier borne weight as one of their highest priorities. The IS has addressed this issue directly, exerting extensive effort into minimizing the weight, even at the expense of range. It is paramount to reduce the weight as much as possible to minimize effects on weapon ergonomics (e.g. weapons center of gravity) and human factors. By integrating the LW weapon components into a single weaponsight, significant weight reductions are possible. Some of the IS design decisions responsible for reducing the weight were:

-   Using a shared housing and mounting bracket;
-   Using a CCD Camera instead of Direct View Optics for day operations;
-   Using non-metallic material for the housing;
-   Using a common power supply (one battery versus several) and power conditioner;
-   Using uncooled thermal imager technology that does not use a scanner or cryo-cooler.

The IS has used the lightest subsystem components presently available, and as technology advances in the subsystem areas, further weight reductions are anticipated. Table 6 provides a weight comparison of the IS and LW weapon subsystem. As shown in the table, the complete LW weapon subsystem weighs an additional 2.5 lb. compared to the IS. It should be noted that the LW design allows for selecting components for each mission, which could reduce the weight on the rifle. For instance the CCD camera may be removed at night and the thermal imager removed in the daytime. From an operational standpoint, the modular design provides more flexibility.

COMPONENTS	LW	IS
Thermal Imager (including mount & BA-5847 battery)	2140g	2400g (includes mount, battery, & wiring harness)
Laser Pointer (including mount & batteries)	211g	
LRF/DCA (including mount & battery)	891g	
CCD Camera (including mount)	113g	
Wiring Harness	196g	
TOTAL WEIGHT	3551g (7.8 lb.)	2400g (5.3 lb.)

Table 6: IS and LW weapon subsystem Weight Comparison

POWER

Power consumption is another critical area of concern when designing new soldier equipment. Reducing power consumption lowers the number of batteries required, which equates to lower life cycle costs as well as reduction in the soldier-borne weight. Integration of the subsystems, using common power converters and power management techniques, and use of low-power uncooled thermal imager technology are responsible for the significant power savings of the IS over the LW weapon components. The IS requires only one battery to power all of the subsystems, whereas the LW components require batteries for the thermal imager, LRF/DCA and laser pointer. The LW computer/radio subsystem battery powers the LW CCD Camera. By making power reduction a key focus during the design phase of the Integrated Sight, significant power savings were achieved. Even small reductions in power requirements equate to large savings over the life of a piece of military hardware.

The power consumption of the CCD camera, LRF, electronic compass, and infrared laser pointer for both the IS and LW weapon subsystem are expected to be comparable. (Measured power consumption data for the CCD camera, LRF, electronic compass, and infrared laser pointer was not available at the time this paper was generated.) The main power saving is achieved using the uncooled thermal imager technology in the IS. Table 7 provides the power consumption of the thermal imager in the "Standby" and "On" mode for both systems.

Component	IS	LW
Thermal Imager		
- Standby mode	1.4 W	5.5 W
- On mode	5.8 W	12.0 W

Table 7: Power consumption of the IS and LW Thermal Imager

WEAPON COMPATIBILITY

Due to the modular design of the LW weapon subsystem, missions requiring full LW weapon subsystem capability (i.e. thermal sight, CCD camera, LRF/DCA, and laser pointer) require four mounting rails for mounting of all the components to the weapon. The only weapons that are presently capable of supporting four mounting rails, are the new M16A4 and M4 Modular Weapon Systems (MWS). Both of those weapons are equipped with the Picatinny Rail adapter system and a removable carrying handle. These two weapons contain four rails around the barrel and one above the breech. The LW thermal sight is mounted on the rail above the breech and the other weapon subsystem components are mounted on the rails around the barrel. The LW weapon subsystem components may be used with any weapon that uses the Picatinny Rail as the basis for mounting components. However, only individual LW weapon subsystem components can be mounted on the M240B Machine Gun, the M249 Squad Assault Weapon, and other weapons with only one Picatinny rail. Mounting of components on various weapons will require a different wiring harness for each weapon. Wiring harnesses have been developed to mount some of the LW components on the M240B and M249 weapons.

The IS integral design, eliminates the need for multiple mounting rails, and hence provides full LW weapon subsystem capability, not only on the M16A4 and M4 MWS weapons, but also on other current in-service weapons such as the M16A1/A2, M4, M249, M240 and M60. The IS mounts to the rail above the breech on the M16A4 and M4 MWS. Test fits on each weapon have shown no operational interference with either shell ejection or the charging handle.

While the LW modular approach allows for tailoring for specific missions, it does require several harness configurations. The particular weapon used determines the primary harness configuration. Addition of extension harnesses provides for the modularity of adding the thermal sight, LRF and CCD Camera to a weapon. The basic harness provides for the two-button switch and connectors for additional components as a baseline. The interface between the weapon subsystem and the rest of the LW system occurs at a single point connection.

The IS approach not only requires fewer wiring harnesses (basic harness for interface to the LW system, and two tethers for the LRF and laser pointer remote fire switch), but reduces the overall size and weight, which should improve weapon ergonomics. The reduction in size and weight, and the removal for the need for mounting components on the weapon barrel, should improve the center of gravity of the weapon, with the IS mounted.

BORESIGHTING

For the engagement of targets, boresighting accuracy of the weapon sighting system is a critical parameter. A well-designed weapon sighting system not only achieves accurate boresight alignment, but also ensures boresight retention. Two other important factors to the soldier for sighting systems are the level of tasks and time taken for a soldier to zero the weapon sighting system to the weapon.

One of the key advantages of the IS design over the LW modular design is the simplification of the boresighting process. Boresighting can be separated into factory alignment and field boresighting (weapon zeroing). The IS design approach focused on performing most of the alignment of subsystems at the factory level, hence reducing the weapon zeroing effort required by the soldier. Unlike the LW system, where the soldier is required to weapon zero the thermal imager, CCD Camera, LRF and laser pointer, only the CCD Camera and laser pointer are required to be zeroed to the weapon for the IS. If the soldier with the LW system uses only the CCD camera in the daytime and the thermal imager at night, he still must zero each component as he mounts them at the day/night transition during his mission. For the discussion below, it is assumed that the CCD camera and thermal imager are both mounted and zeroed at the same time.

IS Boresight Procedure

The IS thermal imager, CCD Camera, and LRF are hard mounted, i.e. no mechanical boresight adjustments exist. The only soft mounted component is the IS laser pointer, i.e. mechanical boresight adjustments exist. The thermal imager and CCD Camera mechanical zero is aligned to the rail grabber and LRF transmitter output beam at the factory. Mechanical zero is defined as the line of sight corresponding to the center of the display. Reticle alignment of the thermal imager, CCD Camera and LRF is also performed at the factory. The thermal imager reticle is aligned to the CCD Camera reticle, and the LRF reticle is aligned to both the thermal imager and CCD Camera reticles.

Weapon zeroing is accomplished by the following procedure:

1. Install a laser boresight alignment tool into the weapon barrel, and adjust the CCD Camera reticle position (via the 5-way switch) until the observed beam spot overlays the reticle. Once this is accomplished and saved, not only will the CCD Camera be weapon zeroed, but also the thermal imager and LRF. The thermal imager and LRF reticles will automatically be adjusted to ensure weapon zeroing via the IS software.

2. To boresight the laser pointer, activate the laser pointer and aim it at an object 25 meters away, observe the beam spot via the CCD Camera, and adjust the azimuth and

elevation of the beam spot (via the adjustments on the laser pointer) to overlay the CCD Camera reticle.

LW Boresight Procedure

The LW thermal imager, CCD Camera, LRF and laser pointer are all soft mounted i.e. all have mechanical boresight adjustments. To weapon zero the LW weapon subsystem, weapon zeroing of the thermal imager, CCD camera, LRF and laser pointer needs to be performed individually.

Weapon zero is accomplished by the following procedure:

1. A boresight target is placed 10 meters from the weapon. The boresight target is a grid sheet with alignment marks for the center of boresight target, thermal imager, CCD camera, laser rangefinder and laser pointer aimpoints. A laser boresight alignment tool is installed into the weapon barrel and the observed beam spot is overlaid on to the center of boresight alignment mark.

2. To zero the thermal imager, a heated target is required at the thermal imager alignment mark on the boresight target. A second soldier places a finger on each side of the alignment mark to produce the heated target. The thermal imager reticle is adjusted until the reticle overlays the heated target.

3. Zeroing of the CCD camera is achieved by adjusting the azimuth and elevation of the CCD reticle until it overlays the CCD camera alignment mark on the boresight target.

4. To zero the laser rangefinder, the laser rangefinder built in visible boresight laser is adjusted until the projected spot overlays the laser range finder alignment mark on the boresight target.

5. To zero the laser pointer, the laser pointer is activated, and the azimuth and elevation is adjusted until the projected spot overlays the laser pointer alignment mark on the boresight target.

The IS design has benefited from close focus on boresight retention. Some of the efforts to ensure boresight retention are: minimizing thermal gradients in the chassis, provide adequate structural strength to the chassis, analyze boresight sensitivities to optical shifts, and use proven weapon mounting techniques. However, the greatest anticipated advantage that the IS has compared to the LW weapon subsystem (with respect to boresighting) is the reduced weapon mounting adapters (rail grabbers) required for missions requiring full LW weapon subsystem capability (i.e. thermal sight, CCD Camera, LRF and laser pointer). For full LW weapon subsystem capability, four weapon mounting adapters are required, whereas the IS requires only one weapon mounting adapter, resulting not only in potentially better boresight retention and accuracy alignment, but reduced level of effort and time for weapon zeroing.

M203

One of the problems currently facing M16 and M4 Carbine weapons fitted with an M203 grenade launcher is the considerable elevation that the 40mm grenade requires to achieve its maximum range. The IS program has explored and implemented a solution, to allow the Army to evaluate and consider implementing for future LW systems. Initially, the concept was to use a special superelevation bracket that mounts the sight off to the side of the weapon. The sight is tilted towards the ground, so that when the soldier elevates the weapon to launch a grenade, the sight points towards the target. The superelevation bracket concept had several problems. The most serious problem was that while the sight was attached to the superelevation bracket, it precluded the grenadier from using the sight for the M16 or M4. It is critical that a soldier in combat be able to fire his weapon at any given time. Weapon ergonomics would also be impacted since the weight of the sight mounted off to the side places a rotating moment on the weapon. Boresight retention was also another issue of great concern. The ability for the sight to maintain boresight after repeated elevation of the mounting bracket and weapon firing shock was doubtful.

The IS will demonstrate a method of aiming the M203 which eliminates the need and problems of a special elevation bracket through software, the electronic compass, and a "virtual" reticle. To engage a target with the M203, the soldier ranges to the target using the LRF. The software uses the target range and the output of the electronic compass to determine the line of bearing to the target. Once the M203 ballistic solution is computed, an aiming point is placed on the display along with the weapon reticle. The soldier then elevates the weapon until the weapon reticle is superimposed on the aiming point and launches the grenade. While the weapon is being elevated to its firing position, the IS software uses the output from the accelerometers in the electronic compass to maintain the relative position of the aiming point in the display to the computed firing line of sight needed to engage the target. The advantages that the IS provides is that the grenadier can still use the sight for the M16/M4, boresight retention is maintained, and the requirement for the special elevation bracket is eliminated providing better weapon ergonomics and reduction in weight. The IS also enables the M203 to be fired from the hip or other more comfortable postures when used by a LW equipped soldier via the head mounted display.

The LW weapon subsystem could provide the same capability if required. The LW LRF and electronic compass provides a slew mode to measure the azimuth and vertical angle to compute ballistic solutions and indicate a firing solution through the Soldier Computer Interface. This capability is currently not required by the user.

HUMAN FACTORS

The human factors engineering aspect focuses on optimizing the soldier-machine performance under operational conditions. Parameters such as operating complexity, user comfort or fatigue, number of 'steps' required to perform a function, and time required to perform a function, impact the soldier's lethality and survivability.

Extensive efforts from early in the IS program focused on obtaining optimum implementation of controls and display information, while avoiding unnecessary user complexity and cognitive overload. Much of the soldier's shooting will be "snap and reaction" type shooting, firing that must be done quickly, from unsupported and unexposed (e.g. around corners) firing positions, without the time for complicated weapon preparations. Weight, size, center of gravity location, and complexity of operation are of great importance in engaging and determining location of targets. It is anticipated that the compact size, reduced weight and improved center of gravity of the IS will improve the soldier's ability to aim and fire. Transportability of the IS over long marches will also be less strenuous, reducing the level of soldier fatigue.

A potential area of concern with these systems is that the weapon harnesses could get snagged when a soldier is traversing through dense shrubs or is low crawling along the ground. Because of its fewer harnesses and protrusions, the IS should be less prone to snagging. Boresighting, changing batteries and mounting of weapon subsystems are additional examples of areas where the IS has reduced the level of soldier effort and time taken to perform tasks.

CONTRACT STATUS

Raytheon Texas Instruments Systems has been under contract since June, 1996 to design, build, and deliver twelve IS prototypes. Detailed design is complete and integration began in January 1998. Currently, three IS have been built and are undergoing environmental, performance and reliability testing. The IS prototypes are scheduled to be completed in June 1998.

SUMMARY

The Land Warrior weapon subsystem will provide dismounted infantry soldiers with greatly improved fighting capabilities. It is based on the concept of modular components, which will allow soldiers to tailor their equipment to the planned mission. The Integrated Sight Technology Demonstration will allow the Army to evaluate the concept of integrating the multiple sensors into a single system. The main advantages of integration are reduced weight and power, improved center of gravity, fewer cables and batteries, and easier boresighting and mounting. These improvements should enhance the soldier's lethality and survivability in both direct fire and call for indirect fire situations. The Integrated Sight also provides the US Army with full LW weapon system capability for other in-service weapons, without impact on the weapon ergonomics. The Integrated Sight prototypes are technology demonstrators and it is expected that further weight reductions and performance improvements can be made in the future. However, the prototypes are being designed and tested to allow for thorough field evaluation for potential future insertion into the Land Warrior System.

Soldier Systems Sensor Fusion

Kathryne M. Brubaker
Soldier Systems Command
U.S. Army Natick Research, Development &Engineerging Center
Natick, MA 01760

ABSTRACT

This paper addresses sensor fusion and its applications in emerging Soldier Systems integration and the unique challenges associated with the human platform. Technology that provides the highest operational payoff in a lightweight warrior system must not only have enhanced capabilities, but have low power components resulting in order of magnitude reductions coupled with significant cost reductions. These reductions in power and cost will be achieved through partnership with industry and leveraging of commercial state of the art advancements in microelectronics and power sources.

A new generation of full solution fire control systems (to include temperature, wind, and range sensors) and target acquisition systems will accompany a new generation of individual combat weapons and upgrade existing weapon systems. Advanced lightweight thermal, IR, laser and video sensors will be used for surveillance, target acquisition, imaging and combat identification applications. Multifunctional sensors will provide embedded training features in combat configurations allowing the soldier to "train as he fights" without the traditional cost and weight penalties associated with separate systems.

Personal status monitors (detecting pulse, respiration rate, muscle fatigue, core temperature, etc.) will provide commanders and higher echelons instantaneous medical data. Seamless integration of GPS and dead reckoning (compass and pedometer) and/or inertial sensors will aid navigation and increase position accuracy. Improved sensors and processing capability will provide earlier detection of battlefield hazards such as mines, enemy lasers and NBC (nuclear, biological, chemical) agents. Via the digitized network the situational awareness database will automatically be updated with weapon, medical, position and battlefield hazard data.

Soldier Systems Sensor Fusion will ultimately establish each individual soldier as an individual sensor on the battlefield.

Keywords: Soldier System, Sensor Fusion, Integration, human platform, Land Warrior, FXXI Land Warrior

1. INTRODUCTION

The US Army is on the verge of a revolution in how it will improve the lethality, survivability, command and control, mobility, and sustainment of its units and allow them to accomplish their missions on the digitized battlefield of the 21st century. Military downsizing, budget constraints and technological advances that provide greater unit dispersion, means that our soldiers will be asked to do more with fewer numbers. This resulting "force multiplier" necessitates a creative approach to integrating doctrine, the soldier, and advanced technology. The paramount challenge for Soldier Systems is the integration of these advanced technologies onto the human platform compounded with the other challenges of electronics packaging, EMI and ruggedization of components to withstand a soldier's environment.

2. THE SOLDIER SYSTEM

The Soldier System is generically defined as the individual soldier and everything he/she wears, consumes, or carries for individual use in a tactical environment. Over the past several years the systems approach to modernizing the soldier has been implemented and successfully demonstrated. The current thrust is focused on optimizing the soldier's effectiveness through: 1) the synergy that results from effective integration of technologies at the systems level; and 2) the proper integration of soldier systems across a diverse spectrum of operations. Additionally, the benefits derived from developing the soldier system, like other major weapon systems by applying a systems approach, has accelerated product development cycles, lowered acquisitions costs and reduced requirements. The Soldier Systems journey to the Army After Next's (AAN) Lightweight Soldier (in the year 2020) begins with the soldier of today and his current equipment and technology. The *First*

Part of the SPIE Conference on Sensor Technology for Soldier Systems ● Orlando, Florida ● April 1998
SPIE Vol. 3394 ● 0277-786X/98/$10.00

73

Integrated Soldier System, the Land Warrior (LW) will be fielded by the year 2000. Force XXI Land Warrior (FXXI LW), an S&T program, will demonstrate and integrate technology upgrades to the LW System. These technologies will be included in future productions of LW.

Land Warrior System & FXXI LW Technology Upgrades Overview.

The Computer/Radio Subsystem (CRS) provides a computer based on Intel 486 architecture. It controls all aspects of system operation, control and data handling and provides the soldier with all the information he needs when he needs it via the Soldier Computer Interface (SCI). The soldier and squad radios are based on Motorola commercial technology and provide SINCGARS Interoperability. Added features are GPS, COMSEC, video processing, and a hand held flat panel display. FXXI LW technology upgrades are Enhanced Soldier Radio (ESR), System Voice Control (SVC) and Integrated Navigation (IN). The ESR increases the link margin, which results in increasing the Soldier Radio range to 5 km. The ESR improves communications reliability and for some missions, provides weight reduction because a second radio is not needed. Soldier Radio packet relay protocol will also extend the Soldier Radio range, provide non line-of-sight communications, improve the delivery of data messages in MOUT and other obstructed terrain. SVC, which is a low-complexity, high-accuracy, state-of-the-art voice recognition algorithm, will provide hands-free control of radios, displays and the computer. Hands-free operation enhances a soldier's lethality, mobility, and survivability by enabling him to keep his eyes and weapon down range. IN is based on dead reckoning technology packaged in a module containing a miniature compass and accelerometer. IN provides highly accurate position data with and without GPS satellite information. CRS upgrades will also provide compatibility with the combat identification component.

The Protective Clothing and Individual Equipment Subsystem (PCIES) consists of improved body armor providing improved ballistic protection for the soldier, load carrying vest and pack frame, modular pack, chemical/biological garment, and other clothing and individual equipment. The load carrying assembly provides environmental protection for the computer, radio and cabling.

The Integrated Helmet Assembly Subsystem (IHAS) is based on a reduced weight PASGT helmet and suspension with a COTS headset assembly. The IHAS integrates the helmet, a helmet-mounted monocular day/night sensor and flat panel AMEL display, a laser detection system, ballistic/laser eye protection, and the M45 chemical/biological mask. The FXXI LW technology upgrade for the IHAS subsystem is a Head Orientation Sensor (HOS). The HOS provides Rapid Target Acquisition (RTA) by correlating head orientation with weapon orientation. This results in an increase in target acquisition time and immediate target identification when switching from night vision to weapon sight video.

The Weapon Subsystem (WS) integrates a laser rangefinder, digital compass assembly (LRF/DCA), thermal weapons sight (TWS), AN/PAC-4C IR laser aiming light, close-combat optic and wiring harness with the soldiers primary weapon. A ruggedized video camera is also mounted on the weapon rails. Data/imagery from each is made available to the soldier, the computer and to his command structure by full integration with the other LW subsystems. FXXI LW technology upgrades are an Enhanced Weapon Interface (EWI), an Integrated Sight (IS) and LW-Combat Identification (LW-CID). The EWI will improve mobility and reliability by providing both a wireless link and improved LW cable. The wireless link will eliminate the soldier-to-weapon cable tether. The improved LW cable will weigh less than the current cable and provide improved durability of the cable and connectors. The IS will replace the LRF/DCA, TWS, IR aim light and video camera with an overall weapon weight saving of 4.2 pounds. LW-CID will integrate the functionality of the stand-alone Combat Identification for the Dismounted Soldier (CIDDS) system, which is based on a laser interrogation, RF response query and answer concept. The CIDDS weapon unit will provide combat identification, Tactical Engagement Simulation (TES) capability and an IR aim light. The IR aim light capability will eliminate the AN/PAC-4C and the TES will eliminate the use of a MILES Small Arms Transmitter (SAT) during training exercises. The integration of CIDDS functionality onto the LW platform maximizes the use of existing LW equipment by leveraging the CRS (processing), the Soldier Radio and the helmet laser detectors.

The Software Subsystem integrates tactical and mission support software with the Soldier Computer Interface (SCI). It has been designed to make extensive re-use of NDI software. New software will only be required for LW system integration and the SCI.

3. APPROACH TO FUTURE SOLDIER SYSTEMS

Soldier systems are unique in that they are dominated by the human interface. The main challenge facing soldier systems development is the integration of advanced technologies onto the human platform. Information need must be balanced with the means of conveying that information. In order to address this technical challenge, a truly revolutionary approach to soldier systems development is required; an approach which evolves from a focus on warfighter-centered design and fightability.

3.1 Warfighter-Centered System

A warfighter-centered system by definition is focused on the human element: a true paradigm shift in the way that the system is defined, designed and integrated. In the future, the performance of the soldier, when using a product, will be the measure of success and not the performance of the product. Technology will be driven and designed around the human while recognizing that every human is unique. Unique physically and mentally. Each soldier is coming from his own perspective, based on his own personal experiences. He brings with him his own preferences, abilities and aptitudes.

A warfighter-centered approach to development will require a head-to-toe analysis of all required warfighter components including Government furnished equipment (GFE) items. Every added or improved functionality will be scrutinized by iteratively holding key system drivers as independent variables.

3.2 Key System Drivers

The key system drivers of a warfighter-centered system are weight, power, cost (unit and life cycle), and fightability. Future systems will have established exit criteria for each of these four system parameters. Any increase in weight or power must be accompanied with a verified, measurable increase in the operational payoff for the soldier. And, of course, if the system is unaffordable, the soldier will not reap the benefits of advanced technology.

3.2.1 Weight. The User community has identified system weight reduction as the number one priority for Science and Technology investment. System weight significantly affects a unit's tactical mobility and a soldier's maneuverability, workload and fatigue. The first step in achieving the 50-pound combat load goal will be to perform a rigorous analysis of the current soldiers' equipment load and its capabilities. The correlation between weight and power must not be under emphasized. (i.e., Holding battery technology constant while increasing energy consumption will result in increased battery use, leading to increased logistics and weight burden to the soldier.) The greatest system weight savings will be achieved by using low power, miniaturized electronic components in concert with advanced lightweight materials that provide improved protection and durability.

3.2.2 Power. The next generation warfighter will have substantially increased capabilities - provided in large part by devices requiring electrical power. Yet, individual power was identified by the 1991 Army Science Board as a major barrier to maximizing the soldier's warfighting capabilities. Increased power requirements inherently translates into increased weight and cost. To counter this the following strategies must be pursued in the design and development of all soldier systems: aggressive power management techniques; self-contained, low power devices; and alternate power sources and battery chemistries to include rechargeable batteries. Additionally, the logistic support requirements must be considered taking into account the probability and impact of different combat operations and scenarios, component duty cycles, and the broad range of equipment and vehicles necessary to support the warfighter on the battlefield.

3.2.3 Cost. Both Life Cycle Cost (LCC) and Design To Unit Production Cost (DTUPC) of the system must be considered when assessing system affordability. Historically, the greatest cost "shadow" associated with a system is cast in the production and deployment phases. By implementing Integrated Product and Process Development (IPPD) principles early on in the system design, fewer costly changes will be necessary later. This practice results in a smaller cost shadow being cast during production and over the entire life cycle of the system. IPPD takes into account all system design parameters and specialty engineering disciplines throughout the design process. Parameter examples are reliability and availability of parts, reduction in number of parts, manufacturability, logistics, and producibility. A cost model for LCC and DTUPC will be built using the LW capabilities as a baseline. From this model, cost estimates will be established for future warrior systems. Performing Cost As an Independent Variable (CAIV) analysis on all future warrior technologies will also help ensure system affordability. Further reductions in cost (and power) will be achieved through partnership with industry and leveraging state-of-the-art advances in microelectronics, power sources, software, and computing.

3.2.4 Fightability. To a soldier, optimized human performance can be summarized in one word - FIGHTABILITY. Fightability is necessarily human-centered, yet results from enhanced system performance in the traditional domains of Sustainability, Mobility, Lethality, Survivability, and Command and Control. Fightability has both physical and cognitive components.

Physical Fightability describes a system attribute where the size, shape, weight, integration, and location on the body of individual components serve to maximize the ability of the dismounted warrior to perform the tasks necessary to operate the system in a combat environment. For example, in operational terms, the ability of the soldier to use his weapon without hindrance whenever he needs to is an aspect of physical fightability, as is the ability to use a bayonet effectively with a four-pound sight mounted on the weapon.

Cognitive Fightability describes a system attribute in which the information provided to and the mental demands placed upon the dismounted warrior serve to maximize his ability to perform the tasks necessary to operate the system in a combat environment. For example, in operational terms, presenting the correct amount and quality of information to the soldier, tailored to the combat task that must be performed, is an aspect of cognitive fightability.

Traditionally, products are developed and described in terms of how they must perform. In breaking with tradition, the description of the end product must be re-thought in order to develop system requirements which support fightability as a key system driver. The system must first be described in terms of its physical embodiment, and then in terms of its functions. A methodology that provides a repeatable, quantifiable measurement of fightability will be established by applying an incremental (iterative and cyclical) design approach; utilizing tools such as Quality Functional Deployment (QFD) and Modeling and Simulation (M&S); and performing soldier task analyses, trade studies, warfighter assessments and lab/field evaluations.

The goal is to build dismounted warrior systems that optimize human performance and minimize the effect of individual equipment on the ability of dismounted infantry to perform combat functions.

4. DATA & INFORMATION FUSION

Information drivers for Soldier Systems include: communication, planning, situational awareness, tasking and coordination, fire support, and precision navigation and geolocation. Future soldier systems will have enhanced situational awareness to the extent that every soldier becomes a node of near real time intelligence information while still accomplishing traditional military missions. Through Global Positioning Satellite (GPS) information every soldier becomes an effective land navigator and will have the ability to call for indirect fire to enemy targets and greatly increase the chances of first round hits. The soldier will be able to see farther, in greater detail and have the capability to engage targets farther away with combined arms fire. Digitized, real-time secure communications will ensure the soldier and higher echelons have access to the battlefield information in real-time. Smart systems, whether they are employed in soldier systems, defense surveillance, robotics, or medical diagnosis all require to be aware of subtle details of their environment. That awareness is achieved by interpreting incoming data to construct models relevant to the task. This process is the process of data fusion.

Data fusion is further complicated by data originating from different sources. Each data source must be interpreted differently, and at times, presented distinctly to the soldier. Soldier sensor technology is diverse and soldier systems applications inherently require the fusing of data from multiple sensors. Sensor fusion techniques are increasingly being pursued by the Army and other services. Rapid insertion of this technology into soldier systems is desired.

Sensor fusion performance in solving the soldier's intelligence needs is constrained by the soldier platform. The soldier platform requirements for scalability, processing efficiency, and power as they relate to real time performance have yet to be defined. Meeting those requirements will require significant S&T investment and visibility. With the growth of super computers and the incorporation of high-performance computing (HPC) into user applications challenges will also arise in the presentation of the sensor data.

5. INFORMATION & INPUT/OUTPUT (I/O)

The sheer abundance of sensor data and battlefield information makes the presentation of that information to the soldier more critical. Cognitive fightability will determine the appropriate interface between the human platform and sensor fusion technology. The means of conveying battlefield information are visual, aural, tactile and neuromuscular (e.g., thought recognition or muscular contraction recognition). There is currently no plan for taste or smell. Within each of these means, several parameters compound the design process. Should the display be handheld, wrist-worn or helmet-mounted? What are the optimum quantity, size, style and placement of icons? How are display concerns such as light security, eye safety, shades of gray, contrast, and EMI being addressed? Does the organization of display screens and information location support perception and cognitive needs under high stress conditions? Should alarms and alerts be visually displayed as flashing icons or in spoken voice (female), buzzes or tones? How many times should an alert be provided and at what interval? What level of personal preferences and choices (i.e., programmability) should be provided?

6. EMERGING TECHNOLOGIES

Investment in sensor fusion for the dismounted soldier will become more critical as the technologies of the future come to fruition. Emerging technologies in displays, sensors and person-portable systems will have a profound impact on the aforementioned key system drivers: weight, power, cost and fightability.

6.1 Target Acquisition

The developments in the soldier's ability to acquire targets will dramatically increase his lethality. The Objective Individual Combat Weapon (OICW) will provide "selectable lethality" capability whereby a single weapon has both traditional munitions and exploding munitions capable of defeating targets in defilade out to 2500 meters. Advances in both communications and image sources will provide automatic target acquisition and handoff to appropriate supporting weapons in all conditions, day/night and through obscurants. Other future capabilities include self-guided personnel projectiles, hands free weapon systems, and individual target recognition and tracking.

6.2 Situational Understanding

Complete fusion of thermal, image intensifiers, geolocation, chemical/biological, etc. sensors will take situational awareness to the next level - situational understanding. Future capabilities will provide a full time link to all available battlefield assets, automatically selected for unit mission and position. Real time data updates will be available to and from higher echelons. Other future technologies include preposition sensors, holographic images/decoys, Wide Band Mobile Internet, Manportable Mini-Base Stations, and counter sniper.

6.3 Maneuver

Soldier maneuverability will be increased by dramatically reducing the overall system weight to 50 pounds (goal) for the fighting load. This 50-pound system will have even more capability than the current systems allowing soldiers to move more accurately and quickly. New capabilities include in-stride mine avoidance, automatic selection of the best route, prediction of enemy movements and counter maneuvers. Movement rates will increase with the use of mechanical assist or individual/group transporters and exact airdrop insertions for troops/units.

6.4 Warrior Protection

 Superior force protection will be achieved by fully integrating protective materials, signature control (Near/Mid/Far IR, visual, acoustic, radar), advanced physiological monitoring (muscle response/fatigue, vital organ functions, etc.) and combat identification systems that can distinguish among friendly, enemy or noncombatants. Integrated advanced materials will provide protection from direct fire munitions to head and torso at reduced bulk, logistics and weight of current technology. Novel barrier technologies such as spray-on second skin will provide camouflage and/or NBC protection.

6.5 Sustained Operations

Leap-ahead capability in Manportable Power Sources will provide seven days of sustained operations without re-supply. Alternate compact power sources (microturbines, etc) combined with self-contained, low power devices, and renewable power sources will provide a revolutionary increase in manportable power capability.

7. SUMMARY

Advances in soldier sensor technology will necessitate more investment in soldier platform sensor fusion programs. The number and variety of sensors that will emerge will be constrained by the ability to make sense of their data and present it in a human centered manner. Data analysis must be thorough before presentation, yet allow the soldier access to the raw data unfiltered by software algorithms, for a singular data point to be lost forever. Sensor miniaturization is desirable, but a footprint optimized for human body is imperative. The fusion of audio, visual, environmental, and physiological sensors operating across the electromagnetic spectrum, requiring power and processing and dominated by the human interface poses quite a challenge; the challenge of Soldier Systems Sensor Fusion.

8. ACKNOWLEDGEMENTS

The author wishes to acknowledge FXXI LW Future Warrior Architecture (FWA) Team for their forward thinking approach to future soldier systems development. Their innovative ideas were incorporated in this paper.

9. REFERENCES

1. Proposal for the Land Warrior System, DAAB07-95-R-H302, (March 1995)
2. Future Warrior Architecture Overview Briefing (R. O'Brien; FWA Team Leader, FXXI LW, Natick RDEC)
3. FXXI LW March 1997 In Process Review Briefing Material (L. Pensotti; Arthur D. Little)
4. Army Science and Technology Master Plan, 1997

Further author information –
Email: kbrubake@natick–emh2.army.mil; Telephone; 508-233-5313; Fax: 508-233-4483

Man-Portable Networked Sensor System

W. D. Bryan, H. G. Nguyen, and D. W. Gage

Space and Naval Warfare Systems Center, San Diego
San Diego, California 92152 USA

ABSTRACT

The Man-Portable Networked Sensor System (MPNSS), with its baseline sensor suite of a pan/tilt unit with video and FLIR cameras and laser rangefinder, functions in a distributed network of remote sensing packages and control stations designed to provide a rapidly deployable, extended-range surveillance capability for a wide variety of security operations and other tactical missions.

While first developed as a man-portable prototype, these sensor packages can also be deployed on UGVs and UAVs, and a copy of this package been demonstrated flying on the Sikorsky Cypher VTOL UAV in counterdrug and MOUT scenarios.

The system makes maximum use of COTS components for sensing, processing, and communications, and of both established and emerging standard communications networking protocols and system integration techniques. This paper will discuss the technical issues involved in: (1) system integration using COTS components and emerging bus standards, (2) flexible networking for a scalable system, and (3) the human interface designed to maximize information presentation to the warfighter in battle situations.

Key words: security sensor, integration, human interface, network

1. INTRODUCTION

In today's environment of Operations Other Than War, military personnel are increasingly called upon to enter urban or similarly complex hostile environments and encounter life-threatening situations. The typical high risk conditions the warfighters are exposed to include clearing buildings, neutralizing snipers, and rescuing hostages while avoiding ambush.

The Man-Portable Network Sensor System (MPNSS) has been developed to reduce risk to military personnel by improving their situational awareness and increasing squad efficiency in executing these tactical mission tasks [1]. It consists of a network of portable sensors that can be placed by the soldiers at strategic locations to provide continuous surveillance of critical areas (Figure 1). Still images and continuous video (in both visible light and infrared), motion and acoustic alerts, and range to targets are available to remote control/display stations linked to the radio-frequency (RF) network. Control/display stations can be portable laptop computers, handheld terminal units, soldier-wearable computers or vehicle-mounted computers.

MPNSS has roots in the Air Mobile Ground Security and Surveillance System (AMGSSS) project [2,3], started in fiscal year 1992. The AMGSSS project's goal was to provide a rapidly deployable, extended-range surveillance capability for force protection and tactical security. It consisted of three vertical-takeoff-and-landing (VTOL) air-mobile remote sensing platforms (Figure 2) and a control station. The VTOL air-mobile platform's function is to carry its sensor payload to remote ground locations, up to 10 km from the base station, for extended surveillance missions. These ground locations can be hill tops, roof tops, or any place with clear view of critical areas. This program was supported by the US Army Product Manager-Physical Security Equipment (PM-PSE) and the Office of the Undersecretary of Defense (Acquisition, Tactical Systems/Land Systems).

In 1996 AMGSSS acquired additional support from the US Army Military Police School and the US Army Infantry Center Dismounted Battlespace Battle Laboratory, and changed its name to the Multipurpose Security and Surveillance

Part of the SPIE Conference on Sensor Technology for Soldier Systems ● Orlando, Florida ● April 1998
SPIE Vol. 3394 ● 0277-786X/98/$10.00

79

Figure 1. Artist's concept of the MPNSS ruggedized sensor package.

Mission Platform (MSSMP) [4,5]. The system's role expanded to include new mission areas such as: support to counterdrug and border patrol operations, detection and assessment of barriers (i.e., mine fields, tank traps), remote assessment of suspected contaminated areas (i.e., chemical, biological, and nuclear), and the use of the air-mobile platform as signal/communications relays and to resupply small quantities of critical items.

In 1997, two MSSMP-related Technology Base efforts supported by the Defense Special Weapons Agency were initiated. The first project deals with issues relevant to tactical security sensor internetting and integration [6,7]. One focus is on the identification or definition of Internet Protocol (IP)-based protocols tuned to the requirement for message-based traffic over a heterogeneous RF network whose throughput and delay characteristics are likely to vary in both space and time. Such a protocol must, unlike the Transmission Control Protocol (TCP), provide real-time communications status information in order to allow both the operator and the remote sensor package to adapt their behavior to changing communications capacities in real time. A second focus is on identifying commercial-off-the-shelf (COTS) distributed control techniques that can be used to expedite the effective integration of COTS sensor and other subsystems that provide only slow, asynchronous, character-based RS-232 control interfaces.

Figure 2. The AMGSSS/MSSMP air-mobile platform and integrated sensor pod.

The second project is investigating more advanced human interface techniques for presenting tactical sensor information intuitively and intelligently to the warfighter, preventing information overload during stressful conditions. Topics that are being studied include:

- Alternative input mechanisms (e.g., voice, gesture)
- Alternative displays and cueing mechanisms (e.g., see-through displays, audio cueing)
- More efficient, direct graphic presentation

The findings of this second project, together with experience gained with the portable prototype sensor package during field testing of the AMGSSS/MSSMP system at Military-Operations-Urban-Terrain (MOUT) facilities [5], led us to realize the importance of a man-portable networked sensor system. Without the cost of the air-mobile platforms, a network of man-portable sensor units can be inexpensively produced that will make significant contributions to urban terrain operations. Funding has been obtained to initiate an effort to produce a ruggedized sensor package and incorporate improvements which came out of the AMGSSS/MSSMP system tests.

MPNSS inherits from its predecessors not only a baseline prototype sensor package, but also the following important features:

- The use commercial-off-the-shelf (COTS) sensors and standards-based hardware and software
- A flexible Internet Protocol (IP)-based networked communications architecture

The incorporation of these features enabled the rapid prototyping of a flexible system that can be readily expanded or modified [3].

2. SENSOR SYSTEM

The MPNSS system consists of a scalable suite of smart sensors configured into man-portable packages. Multiple sensor packages are distributed throughout a surveillance area and are linked together by a wireless network communications architecture. Sensor/control data from/to these sensor nodes is available to users throughout this network.

The majority of sensory data processing is performed at the remote sensor node. This reduces both the bandwidth and the power consumption required to transmit the information. Decisions are made by the remote computers so that only useful data is transmitted. Communications only occur when there is information to communicate.

The system architecture is flexible enough to allow easy integration of future sensors and the replacement of current sensors with more advanced models. This architecture provides for scalability: a given sensor or network of sensors can be easily configured to satisfy the mission functional requirements. A sensor package can consist of a single sensor such as a helmet-mounted video camera, an unattended ground sensor, or a suite of integrated sensors. The baseline sensor suite (Figure 3a) provides the following functionality: video/images for visual target classification, image enhancement, motion detection and acoustic alerts, acoustic classification, sensor/image processing for bandwidth reduction, and GPS/heading/range data processing for absolute target positioning.

2.1 MPNSS Baseline Sensors

The baseline MPNSS sensor suite (Figure 3a) includes [5]:

- Visible light video camera
- Infrared video camera (FLIR)
- Laser rangefinder
- Global Positioning System (GPS) receiver and heading sensor
- Interface for an acoustic sensor
- Interface for legacy sensor gateway

Sensor selections were guided by the criteria of minimum power, size and weight, as well as adequacy of performance, low cost and off-the-shelf availability at integration time.

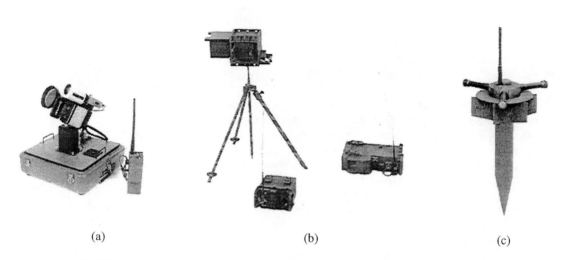

(a)　　　　　　　　　　　　　(b)　　　　　　　　　　　　　(c)

Figure 3. Tested MPNSS sensors: (a) Integrated sensor package, (b) Tactical Remote Sensor System (TRSS) ground sensor, (c) Northrop Grumman's Unattended Ground Sensor. In addition, the MPNSS sensors can also be mounted on VTOL unmanned air vehicles (UAVs) and unmanned ground vehicles (UGVs), as shown in Figure 4.

Visible-Light Video Camera:

A Cohu camera with a Canon 10:1 zoom lens and a 2X range extender was selected. Design specifications called for monochrome images with sufficient resolution to allow classification of vehicles at 2.5 km and personnel at 1 km. This setup allows us to meet the Johnson Criteria [8] for resolution, as well as the sensitivity, dynamic range and signal-to-noise ratio necessary for reliable target detection and classification at those distances.

Infrared Video Camera:

The thermal imager is required to meet the same requirements as the visible-light imaging system but in total darkness. An Inframetrics InfraCam was selected, with a 100mm lens. This imager uses a platinum silicide focal plane array for high pixel uniformity and a proprietary Stirling-cycle dewar cooler to combine light weight with low power and reasonably high image quality.

Laser Rangefinder:

To determine target range, a laser rangefinder was included in the sensor suite. Minimum MPNSS requirement for a laser rangefinder is the ability to reliably determine the range of typical military targets at up to 2,500 meters with 10 meters accuracy. The unit must also be class-1 eyesafe and remotely controllable.

Nine laser rangefinders were found which met the requirements to varying degrees. The Melios, in wide use in the US Army, would probably be the best choice for a fieldable system. However, to save money and weight, we compromised and used a Riegl Lasertape for the prototype system. The Lasertape works well to about 800 meters, sufficient for many applications.

GPS/Heading:

Target position data is determined using a combination of sensor data including GPS, heading, and range. Rockwell's Zodiac sensor, a 12 channel GPS receiver, provides sensor package position data. This sensor comes with an additional port for differential correction updates for improved position precision. Differential updates are transmitted using the existing RF communications network. A heading sensor from Precision Navigation provides target bearing data corrected for sensor tilt and magnetic distortions. The tilt sensor is also used to monitor for disturbance of deployed sensor package. Onboard processing of this data along with range data provides accurate position information of surveillance targets.

Acoustic Detector:

Because of the immaturity and rapid evolution of product development of acoustic detectors, we decided to provide an external data connection for an optional acoustic sensor instead of physically integrating one into the sensor package. A prototype air-deployed acoustic sensor made by Northrop-Grumman (Figure 3c) was connected to the MSSMP Mission Payload Prototype and tested in the field in late 1995. The acoustic package is a small ring of three microphones with custom processing hardware and a serial interface. Output indicates target azimuth angle, type (ground vehicle, jet, helicopter, etc.), and detection/ classification confidence. MPNSS software provides programmable filters that allow the user to discard specified types of sound from specified azimuthal areas.

Legacy Sensor System Gateway:

Legacy sensor systems fielded in the services, such as the Tactical Remote Sensor System (TRSS), Improved Remotely Monitored Battlefield Sensor System (IREMBASS), and Tactical Automated Security System (TASS), employ a standard communications protocol (SEIWG-05) for transmitting sensor data over very-high-frequency (VHF) point-to-point radio links. A small VHF transceiver is embedded in the MPNSS baseline sensor package delivering this sensor data to the payload processor. Software has been written to monitor alerts from these sensor systems and reformat the data into IP packets. This data is multiplexed along with other sensor data and distributed across the network to all users.

A more detailed description of MPNSS sensors can be found in [9].

2.2 Sensor Processing Architecture

Three microcomputers reside in the MPNSS baseline sensor package, processing and reducing the raw data collected before transmitting them back to the Control/Display unit. These take the form of low-power PC/104 form-factor processor boards, running the MS-DOS and MS-Windows 95 operating systems. The three payload computers are interconnected using an Ethernet link for inter-processor communications. Each computer has its own Internet address, and theoretically could be thousands of miles apart, connected by any available IP-based infrastructure, wired or wireless. This architecture has allowed parallel development of the subsystems at different sites, easy debugging by substitution of any processor by an equivalent desktop computer, and has produced a flexible system that can be readily expanded or modified [3].

The Payload Processor handles the communication with the Control/Display, interpreting and executing high level Control/Display commands by generating and sending sequences of simple low-level commands to the various sensor subsystems via RS-232 and other interfaces. It also coordinates the flow of information between all payload computers; monitors, filters, and consolidates alerts received from the Image Processor, acoustic detector, and legacy sensor gateway; and periodically sends status updates for all sensors to the Control/Display unit.

The Image Processor compresses images before sending them back to the Control/Display, using JPEG compression, and performs scene video motion detection when commanded. Image enhancement techniques (contrast enhancement, histogram equalization, etc.), when selected by the operator, are also performed by the Image Processor on the captured image before compression and transmission.

The Video Processor is dedicated to real-time video compressing and transmission. A commercial Windows-95 CUSeeMe software package is used to perform this function. This package features multicast capability which provides IP-based voice and video to multiple users operating within the surveillance network. Currently, the video and voice functions are being upgraded to an embedded hardware codec card which uses H.320/H.323 compression algorithms for improved performance.

The baseline MPNSS package has taken advantage of recent advances in COTS processor performance and component integration driven by the PC notebook computer market. This has resulted in an increase in sensor performance while reducing sensor package power consumption, size, and weight.

3. COMMUNICATIONS NETWORK

The MPNSS communications architecture consists of an all-digital fully-connected adaptive network (Figure 4) providing integrated video, voice, and data services. The fundamental component of this architecture is its basis on the IP suite. This enables plug-and-play compatibility with existing wired and wireless commercial/government off-the-shelf (COTS/GOTS) communications hardware, including evolving Army tactical communications architectures and systems, such as Tactical Internet (TI), Near-Term Digital Radio (NTDR), Joint Tactical Radio System (JTRS), and Warfighter Information Network (WIN). Operational tests this year will demonstrate a mix of COTS/GOTS wireless systems including the Army's NTDR tactical radio, SEIWG-05 radios, COTS 802.11 wireless local area network (WLAN) radios, and Cellular Digital Packet Data (CDPD) modems. Features of these tests will include: (1) on-the-move connectivity from sensor to soldier, soldier to soldier, and sensor to tactical operation centers; (2) integration of legacy sensor systems (TRSS, IREMBASS) through network gateway hardware; and (3) beyond-line-of-sight (BLOS) sensor access/control.

The communications architecture was initially developed using SINCGARS and PRC-139 tactical radios [5,10]. The SINCGARS radio is the Army's projected least common denominator Combat Net Radio. Magnavox/SAIC PCMCIA Tactical Communication Interface Modules (TCIMs), with software modified by us, provided the physical layer for the IP-based network connectivity in accordance with MIL-188-220A over SINCGARS radios. Operational field tests were conducted in 1995, and beyond-line-of-sight network connectivity was established with this tactical IP data link. However, effective data throughput was low (< 500 bps), and COTS WLAN technology was subsequently substituted to improve system performance until military communications technology catches up.

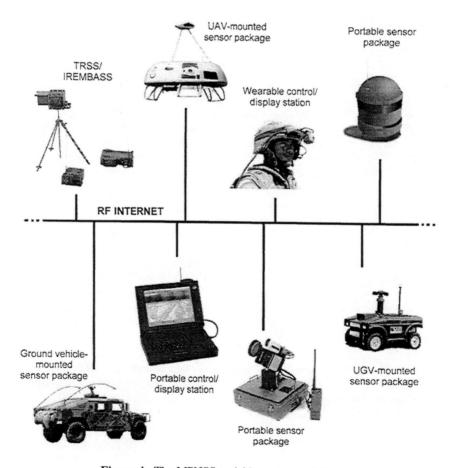

Figure 4. The MPNSS scalable system architecture

The current MPNSS communications architecture provides the following features:

Beyond-line-of-sight communications:

BLOS communications is provided through a combination of multiple line-of-sight (LOS) network communication segments. The IP-based radios self-organize into a tree structured network topology. Each radio maintains a dynamic configuration table of radio node connections within the communications network. IP devices connected to each radio are also listed in the configuration table. This table is used to automatically route data between source and destination nodes. If source and destination nodes are BLOS, the data is routed (hopped) across multiple LOS network segments until it reaches its destination node. If a radio node goes down, radio configuration tables are updated and the data packets passing through the down radio node are rerouted automatically. The Spanning Tree Protocol (IEEE 802.1d) is used to prevent data routing loops from developing and reducing available network bandwidth.

Scalability and Mobility:

Radio nodes (associated with sensors, warfighters, and control stations) are automatically added to or deleted from the network configuration table as they enter or leave the network. A mobile node can "roam" through the network and maintain connectivity. The mobile node is automatically connected to another node (static or mobile) in the vicinity of maneuver. The network topology is updated and data traffic is rerouted to arrive at the new mobile node position without the need of operator intervention.

Integrated voice, video, and data services:

Data packets containing video, voice, and sensor data are multiplexed onto the same RF channel. These packets can be transmitted point-to-point (unicast), point-to-multipoint (broadcast), or multipoint-to-multipoint (multicast). Different types of digital data (voice, video, sensor, and control) are converted to IP data packets by application software. Standard IP network software then performs the function of data multiplexing/demultiplexing automatically. This simple scheme of multiplexing services eliminates the requirement for separate communications systems that have traditionally been needed to provide for each of voice communications, video transmission, and sensor command and control. However, latency-critical services like voice can experience degraded performance on congested networks.

IP gateway for military legacy sensor systems:

The US Department of Defense (DoD) currently has several sensor systems in its inventory including the Marine Corps' TRSS, the Army's IREMBASS, and the Air Force's TASS systems. All these systems use a standard radio communication scheme based on SEIWG-05. Each of these systems represents a stovepipe architecture whereby sensor data is available only to a given sensor display station. An IP-based gateway is being developed to enable DoD legacy sensor data to be accessed by multiple users over a widely distributed network. This wireless gateway reformats the point-to-point legacy sensor data into IP packets which can be distributed horizontally across the MPNSS communications network and vertically to higher echelon commands using TI and WIN.

Flexible mix of existing communications channels:

The IP-based network architecture takes advantage of the vast communication infrastructures, both wired and wireless, that have been developed to pass IP data traffic. MPNSS has been seamlessly demonstrated using a heterogeneous set of communications hardware including tactical radios, COTS WLAN radios, the Internet, Cellular, and Plain Old Telephone System (POTS). Communications tests this year will consists of a mix of COTS/GOTS wireless systems including the Army's NTDR tactical radio, SEIWG-05 radios, 802.11 WLAN radios, and CDPD modems. This flexible mix of communication links and the existing Internet infrastructure enables an extremely wide and flexible network, allowing remote monitoring or control of field sensors by command echelons over vast distances.

Current network communication issues being investigated [6,7] are robust transport protocol development for mobile wireless networks, efficient mechanisms for network data distribution, and intelligent arbitration techniques for control of a sensor by multiple users in a network.

4. HUMAN INTERFACE

The user interface for the MPNSS is a Microsoft Windows-based software program with standard windows, menus, buttons, and dialog boxes. Using a keyboard and mouse, the operator can display remote sensor data, command remote sensors to perform elementary functions--including taking target snapshots (at specified zoom, focus, gain, polarity, azimuth and elevation, etc.), enhancing target images prior to transmission, measuring target range/position, initiating live compressed video of target--or program complex sequences of commands such as composing image panoramas, acoustic filters, and motion detection at various critical scene points. An in-depth discussion of how these commands are executed can be found in [9]. Figure 5 shows examples of two Control/Display screens that provide situational awareness to the operator (details of the Control/Display features were presented in [3]).

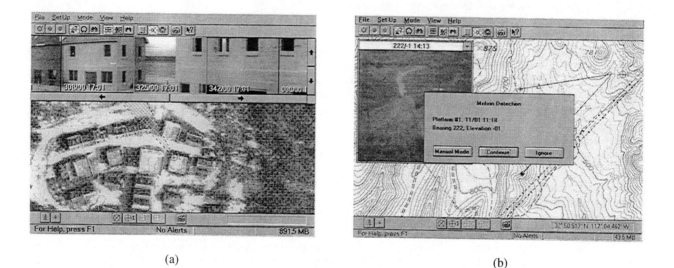

(a)   (b)

Figure 5. Two sample Control/Display screens: (a) Ft. Benning panorama and corresponding aerial photo, (b) Motion detection alert, target image and location on map.

While the Control/Display program was first developed for laptop computers, we have also successfully demonstrated it on a pen computer, a Xybernaut soldier-wearable computer, and the Litton Handheld Terminal Unit (one of the prototype Force XXI Dismounted Soldier System Units) [5] (see Figure 6). The wearable computer tested with the MPNSS system includes a small computer (with integrated trackball) and battery pack that can be worn around the waist, and a small head-mounted display (over one eye) with integrated camera, microphone and earphone. This configuration added several demonstrated capabilities to the system:

- The warfighter can access, monitor, and control remote sensor data while on-the-move.
- Voice control and feedback of sensor system functions provides hands free operation.
- Heads-up display provides viewing of remote sensor data within the warfighter's field of view.
- Video from the head-mounted camera enables the warfighter to function as a sensor within the network.
- Integrated voice communications within the communications network architecture.

Current research efforts are underway to develop more intuitive methods for presenting/controlling tactical sensor data to/by the warfighter that will improve his situational awareness without adding additional stress while performing tactical mission scenarios. This research specifically addresses new metaphors of input/output for the human-computer interface. Input metaphors being investigated include voice, gesture, and touch. Output metaphors include augmented display technology, and retinal, symbolic, audio, and tactile feedback. Wearable and vehicle computer hardware technologies are being investigated for both mounted and dismounted tactical scenarios.

(a)   (b)   (c)

Figure 6. Some tested Control/Display configurations: (a) laptop computer,
(b) wearable computer, (c) Dismounted Soldier System Unit.

5. CONCLUSION

The Man-Portable Networked Sensor System is a flexible, readily extendable network of portable sensors designed to provide improved situation awareness, security, and surveillance capabilities to the warfighter. Sensors may include any combination of the portable baseline integrated sensor packages (with COTS video, infrared, and laser-rangefinding sensors), to new prototype or DoD legacy acoustic ground sensors. The IP based network takes advantage of the myriad of communications channels that have been developed for the Internet, providing a flexible network that allows command, control and monitoring of field sensors by warfighters and command echelons over vast distances.

In January 1997, the prototype MPNSS baseline sensor package was demonstrated (in the same wireless network with its UAV-mounted counterpart) at the US Army's MOUT facility at Ft. Benning, Georgia [5]. The system demonstrated reconnaissance support for advancing troops and security surveillance in urban terrain. In August 1997, the MPNSS portable baseline sensor package was operated in the stand-alone configuration during MOUT exercises at Camp Pendleton Marine Corps Base in California.

Also in 1997, at the request of Naval Air Station Keflavik, Iceland, we developed test plans to demonstrate the long-term, severe-weather use of MPNSS for augmenting existing base security measures. MPNSS sensors would remotely guard aircraft bunkers, flightline, and base perimeter, reducing manning requirements that are costly in the often hostile Icelandic weather. The wireless MPNSS network provides the flexibility needed to easily extend the base security perimeter outward during periods of heightened threat conditions.

Together with on-going technical-base efforts, funded by the Defense Special Weapons Agency, to develop advanced human interface and networking technologies, work has begun on ruggedizing the MPNSS integrated sensor package in anticipation of the tests in Iceland.

REFERENCES

1. "Man-Portable Networked Sensor System," WWW URL: *http://marlin.spawar.navy.mil/D37/D371/mpnss/mpnss.html.*
2. D. W. Murphy and J. P. Bott, "On the Lookout: The Air Mobile Ground Security and Surveillance System (AMGSSS) Has Arrived," *Unmanned Systems,* Vol. 13, No. 4, Fall 1995.
3. H. G. Nguyen, W. C. Marsh, and W. D. Bryan, "Virtual Systems: Aspects of the Air-Mobile Ground Security and Surveillance System Prototype," *Unmanned Systems,* Vol. 14, No. 1, Winter 1996.

4. "Multipurpose Security and Surveillance Mission Platform," *http://www.spawar.navy.mil/robots/air/amgsss/mssmp.html.*

5. D. W. Murphy, J. P. Bott, W. D. Bryan, J. L. Coleman, D. W. Gage, H. G. Nguyen, and M. P. Cheatham, "MSSMP: No Place to Hide," *AUVSI '97 Proceedings,* pp. 281-290, Baltimore, MD, 3-6 June, 1997.

6. D. W. Gage, "Network Protocols for Mobile Robot Systems," *SPIE Proc. 3210: Mobile Robots XII,* Pittsburgh, PA, 14-17 October, 1997.

7. D. W. Gage, W. D. Bryan, and H. G. Nguyen, "Internetting Tactical Security Sensor Systems," *SPIE Vol. 3393: Digitization of the Battlespace,* Orlando, FL, 15-17 April 1998.

8. J. Johnson, "Analysis of Image Forming Systems," *Proc. Image Intensifier Symposium,* US Army Engineer Research and Development Laboratories, Fort Belvoir, VA, 6-7 Oct. 1958 (AD220160).

9. D. W. Murphy et al., "Air-Mobile Ground Security and Surveillance System (AMGSSS) Project Summary Report," Technical Document 2914, NCCOSC RDT&E Div., San Diego, CA, September 1996 (AD-A317618).

10. B. Martin and D. Bryan, "Low-Cost Miniature Interface and Control Systems for Smart Sensors, Tactical Radios, and Computer Networks," IEEE Military Communications Conference (MILCOM 95), San Diego, CA, Nov. 6-8, 1995.

Addendum

The following papers were announced for publication in this proceedings but have been withdrawn or are unavailable.

[3394-01]   **Sensors for soldier systems: an overview**
P. R. Snow, Jr., U.S. Army Natick Research, Development and Engineering Ctr.

[3394-04]   **Flight assessment of near-term night-vision goggle technology for helicopter pilotage**
P. M. Steele, U.S. Army Night Vision & Electronic Sensors Directorate and QuesTech Inc.; J. B. Gillespie, U.S. Army Night Vision & Electronic Sensors Directorate

Author Index